ANALYTICAL CHEMISTRY SYMPOSIA SERIES — volume 16

new approaches in liquid chromatography

Proceedings of the 2nd Annual American–Eastern European Symposium on Advances in Liquid Chromatography, Szeged, Hungary, June 16–18, 1982

ANALYTICAL CHEMISTRY SYMPOSIA SERIES

Vol. 1 Recent Developments in Chromatography and Electrophoresis. Proceedings of the 9th International Symposium on Chromatography and Electrophoresis, Riva del Garda, May, 15–17, 1978
edited by A. Frigerio and L. Renoz

Vol. 2 Electroanalysis in Hygiene, Environmental, Clinical and Pharmaceutical Chemistry. Proceedings of a Conference, organised by the Electroanalytical Group of the Chemical Society, London, held at Chelsea College, University of London, April 17–20, 1979
edited by W. F. Smyth

Vol. 3 Recent Developments in Chromatography and Electrophoresis, 10. Proceedings of the 10th International Symposium on Chromatography and Electrophoresis, Venice, June 19–20, 1979
edited by A. Frigerio and M. McCamish

Vol. 4 Recent Developments in Mass Spectrometry in Biochemistry and Medicine, 6. Proceedings of the 6th International Symposium on Mass Spectrometry in Biochemistry and Medicine, Venice, June 21–22, 1979
edited by A. Frigerio and M. McCamish

Vol. 5 Biochemical and Biological Applications of Isotachophoresis. Proceedings of the First International Symposium, Baconfoy, May 4–5, 1979
edited by A. Adam and C. Schots

Vol. 6 Analytical Isotachophoresis. Proceedings of the 2nd International Symposium on Isotachophoresis, Eindhoven, September 9–11, 1980
edited by F. M. Everaerts

Vol. 7 Recent Developments in Mass Spectrometry in Biochemistry, Medicine and Environmental Research, 7. Proceedings of the 7th International Symposium on Mass Spectrometry in Biochemistry, Medicine and Environmental Research, Milan, June 16–18, 1980
edited by A. Frigerio

Vol. 8 Ion-Selective Electrodes, 3. Proceedings of the Third Symposium, Mátrafüred, Hungary, October 13–15, 1980
edited by E. Pungor and I. Buzás

Vol. 9 Affinity Chromatography and Related Techniques. Theoretical Aspects — Industrial and Biomedical Applications. Proceedings of the 4th International Symposium Veldhoven, The Netherlands, June 22–26, 1981
edited by T. C. J. Gribnau, J. Visser and R. J. F. Nivard

Vol. 10 Advances in Steroid Analysis. Proceedings on the Analysis of Steroids, Eger, Hungary, May 20–22, 1981
edited by S. Görög

Vol. 11 Stable Isotopes. Proceedings of the 4th International Conference, Jülich, March 23–26, 1981
edited by H.-L. Schmidt, H. Förstel and K. Heinziger

Vol. 12 Recent Developments in Mass Spectrometry in Biochemistry, Medicine and Environmental Research, 8. Proceedings of the 8th International Symposium on Mass Spectrometry in Biochemistry, Medicine and Environmental Research, Venice, June 18–19, 1981
edited by A. Frigerio

Vol. 13 Chromatography in Biochemistry, Medicine and Environmental Research, 1. Proceedings of the 1st International Symposium of Chromatography in Biochemistry, Medicine and Environmental Research, Venice, June 16–17, 1981
edited by A. Frigerio

Vol. 14 Chromatography and Mass Spectrometry in Biomedical Sciences, 2. Proceedings of the International Conference on Chromatography and Mass Spectrometry in Biomedical Sciences, Bordighera, Italy, June 20–23, 1982
edited by A. Frigerio

Vol. 15 Computer Applications in Chemistry. Proceedings of the 6th International Conference on Computers in Chemical Research and Education (ICCCRE), Washington, DC, July 11–16, 1982
edited by S. R. Heller and R. Potenzone

Vol. 16 New Approaches in Liquid Chromatography. Proceedings of the 2nd Annual American–Eastern European Symposium on Advances in Liquid Chromatography, Szeged, Hungary, June 16–18, 1982
edited by H. Kalász

ANALYTICAL CHEMISTRY SYMPOSIA SERIES — volume 16

new approaches in liquid chromatography

Proceedings of the 2nd Annual American–Eastern European Symposium on Advances in Liquid Chromatography, Szeged, Hungary, June 16–18, 1982

edited by

H. Kalász

Department of Pharmacology
Semmelweis University of Medicine
Budapest, Hungary

ELSEVIER
Amsterdam – Oxford – New York – Tokyo 1984

This volume is published by Akadémiai Kiadó as Vol. 24 in the series Symposia Biologica Hungarica

Joint edition published by
Elsevier Science Publishers, Amsterdam, The Netherlands and Akadémiai Kiadó, The Publishing House of the Hungarian Academy of Sciences, Budapest, Hungary

Distribution of this book is being handled by the following publishers:

for the U.S.A. and Canada
Elsevier Science Publishing Company, Inc.
52 Vanderbilt Avenue
New York, New York 10017

for the East European countries, Democratic People's Republic of Korea, People's Republic of Mongolia, Republic of Cuba, Socialist Republic of Vietnam
Akadémiai Kiadó, The Publishing House of the Hungarian Academy of Sciences, Budapest

for all remaining areas
Elsevier Science Publishers
Molenwerf 1,
P.O. Box 211, 1000 AE Amsterdam, The Netherlands

Library of Congress Cataloging in Publication Data

American–Eastern European Symposium on Advances in
Liquid Chromatography (2nd : 1982 : Szeged, Hungary)
New approaches in liquid chromatography.

(Analytical chemistry symposia series ; v. 16)
Bibliography: p.
Includes index.
1. Liquid chromatography–Congresses. I. Kalász, H.
(Huba), 1941– II. Title. III. Series.
QD79.C454A45 1982 543'.0894 83–14070
ISBN 0–444–99642–7

ISBN 0–444–99642–7 (Vol. 16)
ISBN 0–444–41786–9 (Series)

Printed in Hungary

FOREWORD

In chromatographic works classical definitions can be read. One of the basic books declares for instance that "chromatography operates as a system of three components. A stationary phase, a mobile phase and, last but not least, the sample which is supposed to carry through or along the stationary phase."

I believe that chromatography operates as a system of four components, and the fourth one is the personal contact between people who do the job in particular fields of chromatography. A workshop or symposium might be as important as the work which is (or should be) presented. The personal contact and exchange of thoughts have a most powerful catalytic action on scientific work. The inspiration gained from others may have a basic role in the development of a new technique, but especially in the dissemination of a new method. Chromatography is an extremely rapidly-growing field of methodology. The large amount of new instruments, chemicals, solid phases, tricks, tools and ideas form a puzzling picture which may be clarified by talking and exchanging ideas with people who have already had some personal experience with a given subject, or technique.

A Symposium dealing with "New Approaches in Chromatography" could be just another symposium, were it not for a very special characteristic. And this Symposium had one, namely the very tight connection between East and West. It was the second Symposium in this sense. Scientists coming from East and West, their interesting papers and lively discussions demonstrated the fruitfulness of such meetings. This collection of papers will demonstrate to all who did not participate in this Symposium that there were many important topics of general interest for people working in this field.

We thank all those who prepared their papers for publication, and we hope to meet the participants of the Symposium again, as well as many others who will decide, on reading this volume, to participate in our next meeting.

TIBOR DÉVÉNYI

CONTENTS

CHARACTERIZATION OF LIQUID CHROMATOGRAPHIC STATIONARY PHASES

OPTIMIZATION POSSIBILITIES IN LIQUID COLUMN CHROMATOGRAPHY

THIN-LAYER CHROMATOGRAPHY

ANALYSIS OF AMINO ACIDS

ANALYTICAL AND PREPARATIVE SEPARATION OF PEPTIDES AND PROTEINS

HIGH-PERFORMANCE LIQUID CHROMATOGRAPHY

Proc. of the Symp. on Advances in Liquid Chromatogr., Szeged, Hungary, 1982

SEPARATION OF AMINO-SUBSTITUTED BIOLOGICALLY ACTIVE DERIVATIVES OF BETA-MELANOTROPIN BY REVERSED-PHASE HIGH-PRESSURE LIQUID CHROMATOGRAPHY

J.M. VARGA, L. AIROLDI, G. DAVILA-HUERTA

Department of Dermatology, Yale University School of Medicine, New Haven, Connecticut 06510, USA

SUMMARY

1) Amino-substituted derivatives of β-MSH were synthesized for various purposes such as studies on MSH receptors (FITC-MSH), for the production of thiolated intermediates of β-MSH (SPDP-MSH) and for studying the effects of a novel MSH-toxin conjugate (ouabain MSH).1) A mixture of FITC--β-MSH derivatives was separated by RPLC. The formation of FITC-MSH was monitored by simultaneous UV and fluorescence detection. The major fraction of fluorochrome labeled peptide was resolved as a homogeneous peak on rechromatography. The tyrosinase stimulatory activity of the conjugate was 55% that of the unsubstituted peptide (10^{-8}M).

2) SPDP-MSH conjugates were separated by RP-HPLC by using a multiple acetonitrile gradient. Several biologically active SPDP-β-MSH fractions were obtained and were used for the preparation of thiolated β-MSH as intermediates to link the hormone to MSH-reactive biologically active molecules.

3) Ouabain-β-MSH conjugates were separated by RP-HPLC from unconjugated β-MSH and from oxidized products of ouabain. The formation of reaction products was monitored by using (^{3}H)-ouabain as tracer and measuring the radioactivity of the fractions. The tyrosinase-stimulatory activity of the major ouabain-MSH conjugate was 66% of that of the unsubstituted peptide when measured at 0.1 µM concentration. While free ouabain (1 µM) was highly cytotoxic to murine melanoma, the ouabain-β-MSH conjugate had a potent growth-stimulatory effect during the early phase of cell growth.

INTRODUCTION

Currently, High Pressure Liquid Chromatography (HPLC) is the method of choice for the separation and purification of small and medium-sized (up to 20 residues) peptides and their derivatives (1,2). Reversed-Phase HPLC (RPLC) has been used successfully for the separation of biologically active N-hydroxyphenyl popionyl derivative of a β-melanotropin (β-MSH) fragment (1,3). Subsequently, RPLC was used to monitor the time course of β-MSH iodination and to purify the biologically active mono-^{125}I-β-MSH derivative (4). In addition to the production of biologically active radio-labeled MSH, we have been using RPLC for the purification of biologically active amino-substituted derivatives of the same as reported here.

The amino acid sequence of porcine β-MSH is as shown:

Asp-Glu-Gly-Pro-Tyr-Lys-Met-Glu-His-Phe-Arg-Trp-Gly-Ser-Pro-Pro-Lys-Asp

When the octadecapeptide is derivatized with amino-reactive reagents such as aldehydes, isothiocyanates or N-hydroxysuccinimidylates, the formation of mixture is anticipated, consisting of three mono-substituted derivatives (in the N-terminal amino group or the ε-amino groups of lysine #6 and #17), three disubstituted derivatives and one tri-substituted peptide. In order to be able to use the conjugates for studies on MSH receptors they have to be separated from the unsubstituted peptide, from the reagents and from the biologically inactive reaction products. As we show in this report, RPLC is a convenient and rapid method for the purification of biologically active amino-substituted derivatives of β-MSH.

METHODS

1) FITC-β-MSH

To a solution of 4 mg HPLC-purified β-MSH (4) in 1 ml 0.1M Na_2CO_3, 0.7 mg fluorescein isothiocyanate (FITC, Sigma, St. Louis, MO) in 1 ml 0.1M Na_2CO_3 was added and was allowed to react at room temperature for 2 hr. The clear solution was filtered through a Biogel P2 column (0.8 x 20 cm equilibrated with distilled water). The yellow fraction eluted in the void volume was freeze-dried, then redissolved in 0.4 ml distilled water. Exposure to light was reduced by wrapping the Biogel column and other containers of the reaction mixture in aluminium foil. 10-250 μl aliquots of the solution were injected into the HPLC system which consisted of two Waters Associates M6000A pumps, a Model 6600 Programmer, U6K injector. Waters μBondapack C18 0.3 x 30 cm column, Perkin-Elmer LC-75 UV/Vis detector, Perkin-Elmer fluorescence detector and Houston Instruments Dual Pen chart recorder. A linear 20-25% acetonitrile gradient was used in 5mM ammonium acetate (pH 5.8). Biological activity of the fractions was measured in Cloudman S91 melanoma cells by the Pomerantz tyrosinase assay (5).

2) SPDP-β-MSH

To 200 μg HPLC-purified β-MSH in 40 μl phosphate-buffered saline, ph 7.5 (PBS), a solution of 248 μg N-succinimidyl 3-(2-pyridyl-dithio) propianate (SPDP, Pharmacia Fine Chemicals, Uppsala, Sweden) in 80 μl ethanol was added and the reaction mixture was kept at room temperature for ½ hr. The unreacted reagent was neutralized by adding 488 μg ethanolamine in 80 μl PBS. The reaction mixture was injected into a Beckman Model 334 Gradient Liquid Chromatograph which consisted of 2 Model 110 A pumps, a Model 420 microprocessor/programmer, a Waters C18 μBondapak column a fixed wavelength UV detector and a Kipp & Zonen BD 41 two-pen recorder. An acetonitrile gradient was used containing 0.05% trifluoroacetic acid. The acetonitrile concentration was increased according to the following schedule: 10 → 24% in 22 minutes, → 30% in 3 min, → 40% in 30 min, → 60% in 5 min. 0.5 ml fractions were collected, freeze-dried and assayed for tyrosinase activity (5,6).

3) Ouabain-β-MSH

The procedure we used was similar to the one used for the preparation of daunomycin-MSH (7). To a solution of 10 mg Ouabain (Sigma) containing 5 μCi of (^{3}H)-Ouabain (New England Nuclear, Boston, MA), 5 mg $NaIO_4$ was added in 0.5 ml PBS. The oxidation was allowed to proceed at room temperature for 1 hr. The reaction was terminated by adding glycerol to a 50 mM final concentration. To the oxidized ouabain, 15 mg β-MSH was added in 0.5 ml K_2CO_3 (0.15 M, pH 9.5). Following 1 hr reaction at room temperature, the Schiff base was reduced by adding 0.5 mg $NaBH_4$ in 0.1 ml distilled water. The reduction was completed at 37° for 2 hr. The reaction mixture was passed through a Biogel P2 column (0.8 x 20 cm) equilibrated with distilled water. The radioactive fraction appearing in the void volume was freeze-dried then redissolved and aliquots were injected into the Waters HPLC System described in Section 1. The system was operated in the isocratic mode with 24% acetonitrile in 0.1 M ($NH_4H_2PO_4$ pH 5.8). Radioactivity was measured in a Beckman LS 7000 Liquid Scintillation Counter.

RESULTS and DISCUSSION

1) FITC-β-MSH

Fluorochrome-labeled β-MSH has been used earlier for cell surface labeling and for studies of hormone internalization by melanoma cells (6). In these earlier studies, the FITC-peptide conjugates were separated from the reagents by gel exclusion chromatography. As is shown in Figure 1, the reaction mixture excluded from Biogel P2 contains a small amount of un-

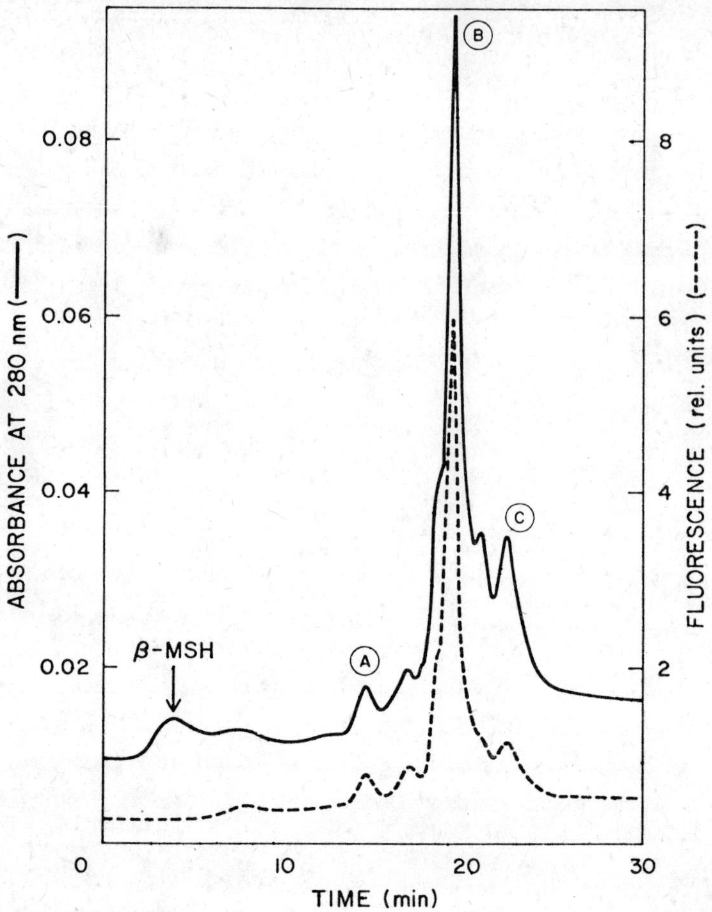

Figure 1 Separation of FITC- β -MSH derivatives, 5 μl FITC- β -MSH solution was injected into the Waters HPLC System at room temperature. Flow rate = 2 ml/min; fluorescence excitation at 490 nm, emission measured at 512 nm.

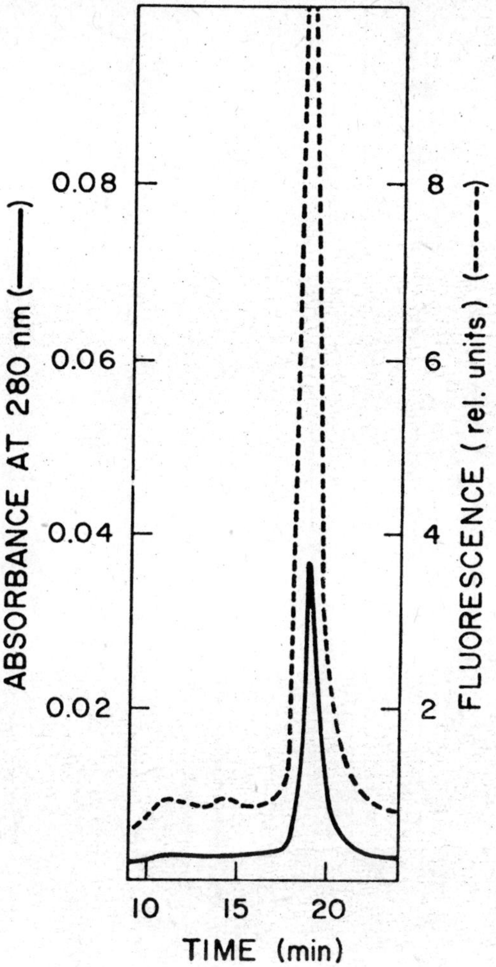

Figure 2 Re-chromatography of FITC-β MSH fraction B (reference to Figure 1). See details is the Methods section. Conditions of HPLC were the same as listed in Figure 1.

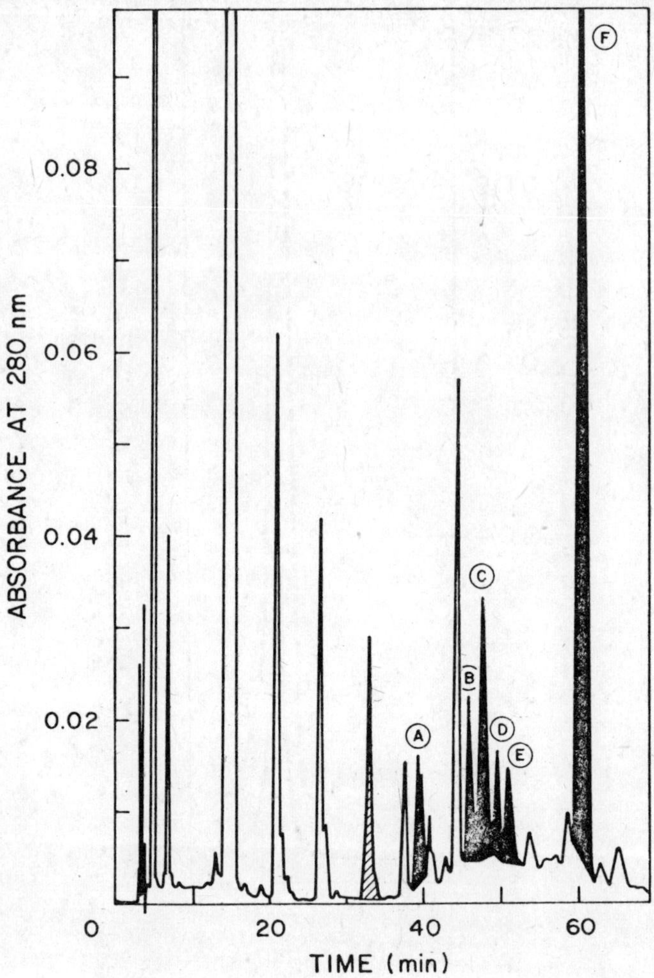

Figure 3 Separation of SPDP- β -MSH derivatives by HPLC. Details of the preparation and gradient are described in the Methods section. Flow rate = 1 ml/min.

labeled peptide. In addition, seven fluorochrome labeled peptide fractions could be detected when the reaction mixture was separated by RP-HPLC. The biological activity of the peaks was measured and compared to the tyrosinase-stimulatory activity of unsubstituted β-MSH (10^{-8}M). The activities of peaks A,B and C were 50, 57 and 23% that of the unsubstituted peptide, respectively. The top fractions of peak B (eluted at 19-20 min) were collected and freeze-dried. The re-dissolved FITC-β-MSH was re-chromatographed under identical conditions. As is shown in Figure 2, the seemingly homogeneous peak could be resolved on re-chromatography. This fraction was collected, freeze-dried and used for studies on MSH receptors. The biological activity at 10^{-8}M concentration was 55% that of the unsubstituted hormone.

2) SPDP-β-MSH

The N-succinimidyl-3-(2-pyridyldithio)-propionyl (SPDP) derivative of β-MSH was prepared in order to introduce a thiol group into the peptide. We used published methods of derivatization with SPDP (8). The separation of the reaction products is an example of the use of RPLC for purifying biologically active hormone derivatives from complex reaction mixtures (Figure 3). In this Figure the unshaded peaks show reaction products that were formed in the reaction of SPDP with β-ethanolamine alone (i.e., with the substance used to neutralize excess SPDP). The line-shaded peak shows the position of unsubstituted β-MSH. The black peaks show the products formed in the reaction of β-MSH and SPDP. When compared with the activity of the unsubstituted peptide at 10^{-8}M, the tyrosinase-stimulatory activity of peaks A-E is in the 50-150% range; the activity of peak F is uncertain. Fractions were subsequently treated with dithiothreitol to reduce the dithio bond and thus form thilated β-MSH which was then used for producing MSH-toxin conjugates. Details of this work will be published later.

3) Ouabain-β-MSH

Ouabain was linked to β-MSH through aldehyde groups formed by periodate oxidation of the deoxymannopyranosyl group of the molecule. Since both the 3-4 and 4-5 carbon bonds of the carbohydrate can be cleaved by periodate oxidation, four different aldehydes can be formed which may react with the three free amino groups of β-MSH. The simplicity of the HPLC elution pattern (Figure 4) may be a consequence of an incomplete resolution of the numerous structurally similar ouabain-β-MSH derivatives; alternatively, few major reaction products could have been formed. Peak A is probably a polymerized product of oxidized ouabain; peak B is native β-MSH; peaks C and D are ouabain-β-MSH conjugates. The tyrosinase-stimulatory activity of fraction (D) is 66% of that of the unsubstituted β-MSH (10^{-7}M). Surprisingly, while free ouabain (1 μM) was highly toxic to melanoma cells, ouabain-β-MSH (peak D) stimulated growth. When ouabain-MSH was added (0.1 μM) to melanoma cell cultures during the first day of cultivation, cells grow about twice as fast as untreated cells. Whether this stimulation is related to an interaction between receptors for MSH and an Na^+/K^+ pump in the cell membrane, is under investigation.

CONCLUSIONS

As we have shown, RPLC is a convenient method for separating and purifying amino substituted derivatives of β-MSH. In some cases well-resolved homogenous fractions can be obtained in one run. In others, re-chromatography is needed to resolve structurally similar biologically active peptide derivatives. On RPLC, amino-substituted derivatives of β-MSH appear later than the unsubstituted peptide. The retardation of the conjugates is probably due to an increased hydrophobicity because the peptide derivatives are formed by the substitution of ionizable amino groups with relatively large aromatic or saturated cyclic compounds (1,9).

ACKNOWLEDGMENTS

We thank Dr. Saul Lande for providing the β-MSH, Noel Richard and Tracy Bobak for their help with the tyrosinase assays, Dr. D. Lambert for setting up the Waters HPLC system, Dr. Gisela Moellmann for correcting and Elena DiMassa for typing the manuscript. This investigation was supported by Grant Number RO1-CA-26081 awarded by the National Cancer Institute, DHEW.

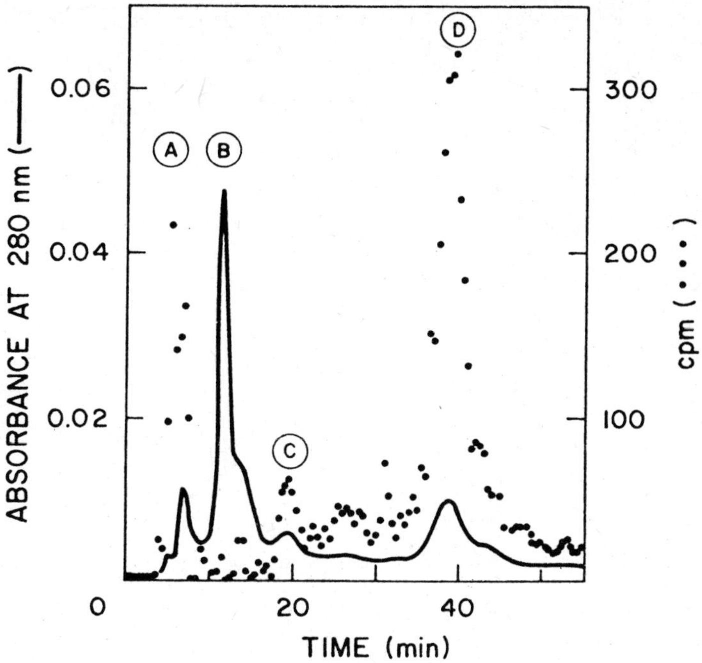

Figure 4 Separation of Oubain- β -MSH derivatives by HPLC. Reaction conditions, apparatus and gradient are described in the Methods section. Flow rate = 2 ml/min.

REFERENCES

(1) MOLNÁR, I. and HORVÁTH, Cs. Separation of amino acids and peptides on non-polar stationary phases by HPLC. J. Chrom. 142: 623-640 (1970)
(2) Proceedings of the International Symposium on HPLC of Proteins and Peptides 1981, Washington, D.C. J.Anal.Biochem. 1982.
(3) ASATO, N., LANDE, S., VARGA, J.M., MOLNAR, I. and HORVATH, Cs. Increase of biological activity of a β-MSH fragment after conjugation with tyrosine-like compounds. Yale J.Biol.Med. 50: 546 (1977)
(4) LAMBERT, D.T., STACHLEK, C., VARGA, J.M. and LERNER, A.B. Iodination of -MSH: time course analysis of reaction mixtures by HPLC and characterization of biologically active mono- and di- iodo β-MSH. J.Biol. Chem. 257: 8211 (1982)
(5) POMERANTZ, S.H. L-tyrosine 3,5 ^{3}H assay for tyrosinase development in skin of newborn hamsters. Science 164: 838 (1969)
(6) VARGA, J.M. and ASATO, N., LANDE, S. and LERNER, A.B. Melanotropin-daunomycin conjugate shows receptor-mediated cytotoxicity in cultured murine melanoma cells. Nature 267: 56-58 (1977)

(7) VARGA,J.M., MOELLMANN,G., FRITSCH,P., GODAWSKA,E. and LERNER,A.B. Association of cell surface receptors for melanotropin with the Golgi region in mouse melanoma cells. Proc.Nat.Acad.Sci.USA 73: 559-562 (1976)

(8) CARLSSON,J., DREVIN,H. and AXEN,R. Protein thiolation and reversible protein-protein conjugation. N-Succin-imidyl 3-(2-Pyridyldithio) propionate a new heterobifunctional reagent. Biochem.J. 173: 723-737, (1978)

(9) MEEK,J.L. Prediction of peptide retention times in high-pressure liquid chromatography on the basis of amino acid composition. Proc.Nat.Acad. Sci. USA 77: 1632-1636 (1980)

Proc. of the Symp. on Advances in Liquid Chromatogr., Szeged, Hungary, 1982

HIGH-PERFORMANCE LIQUID CHROMATOGRAPHY OF FIBROUS PROTEINS

Z. DEYL, K. MACEK

Physiological Institute, Czechoslovak, Academy of Sciences, Prague, Czechoslovakia

In protein separation it is possible to exploit the molecular size, charge differences and the hydrophobicity of species to be separated. According to these properties it is possible to use gel permeation chromatography, ion exchange chromatography or reversed phase HPLC. Each of these techniques has its own advantages and limitations. Thus, for instance, reversed phase chromatography can be used only with proteins which retain their solubility in relatively hydrophobic media and which, if possible, can be recovered in native state after some additional operation. Therefore, reversed phase chromatography has been used mainly with low-molecular weight peptides and proteins and the other two techniques mentioned above are still prevailing in the separations of giant molecules.

A category of far the largest protein molecules is represented by fibrous proteins. Only the category of collagen has been approached by modern chromatographic techniques (1,2) while the others like fibrous proteins of muscle, resilin, fibroin, etc. are at the moment neglected completely. Collagen separations are traditionally a different problem as long as collagen represents a family of very closely related six different proteins of relatively low solutibility that may occur side by side in tissues. From the separation point of view collagens represent a rare category of very high-molecular proteins in which all the three separation principles, i.e., gel permeation, ion exchange and reversed phase partition can be applied. Some of the members of this family (individual collagen types) differ in their relative molecular mass, others in their charge or hydrophobicity, some have a strictly rigid helical structure while in others interruptions of this structure by flexible regions are seen. (For review see. ref.3). Therefore this category of proteins represents an excellent model on which the applicability of different chromatographic principles can be demonstrated. (For review see ref.4).

At the beginning of our studies using HPLC we attempted to separate collagen type I and III polypeptide chains using a copolymer of 2-hydroxy-ethyl methacrylate with ethylene dimethacrylate covalently coated with glucose (Separon HEMA 1000 Glc gel) (Fig.1a, b).

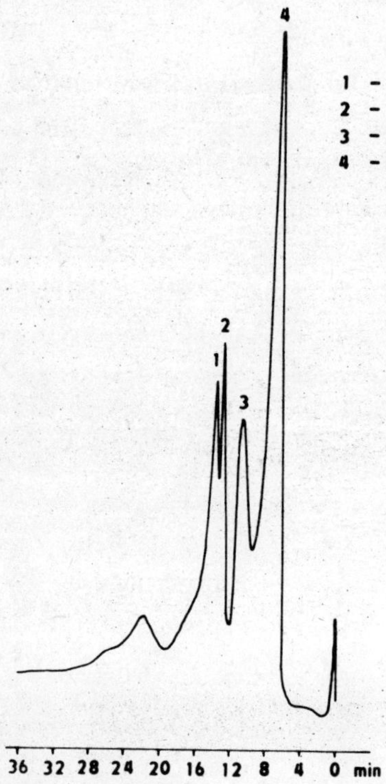

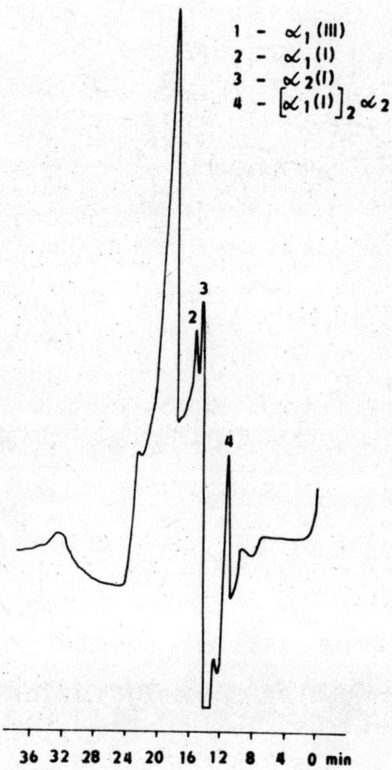

Fig.1a. The chromatographic profile of a mixture of collagen polypeptide chains. Type III collagen is present in its disulphide bonded form $\alpha_1(III)_3$

Fig.1b. The chromatographic profile of a mixture of collagen polypeptide chains. In the mixture disulphide bonds were cleaved by oxidation prior to chromatography and therefore type III collagen is present in its monomeric form as $\alpha 1(III)$

Though a good quality of separations was achieved, the process was not governed solely by gel permeation, as long as molecular entities of identical relative molecular mass were separated. This may be of advantage in the separation of certain collagen mixtures, but causes considerable difficulties when gel permeation separations are used for the investigation of complex mixtures of different collagen polypeptide chains and their fragments. We attempted therefore to abolish the secondary interactions as much as possible and establish a high performance procedure in which gel permeation would be the only mechanism involved.

EXPERIMENTAL

Separations were done on a Pye Unicam liquid chromatograph LC 20 equipped with the UV spectrophotometric detector LC-3 set at 230 nm. A stainless steel column (500 x 8 mm) prepacked with Separon HEMA 1000 Glc, particle size 12-17 µm, was used. The apparatus was operated at about 1·5 mPa overpressure which gave a flow rate of 1.5 ml/min. Of the several mobile phases tested the best resolutions were obtained by isocratic elution with a solution composed of 0.2 mol/L NaCl - 2 mol/L urea - 0.05 mol/L Tris. HCL buffer, pH 7.5. Samples of individual collagen types and their fragments were prepared by established methods. In order to avoid problems arising from UV absorbency of mercaptoethanol, disulfide bonds cleavage was done by performic acid treatment.

RESULTS AND DISCUSSION

The applicability of the worked-out procedure is demonstrated on the following examples. Clearcut separations are obtained with alpha-chain polymers (Fig.2) and a rapid information can be obtained about S-S bond cleavage in collagen type III (Fig.3). The separation efficiency is sufficient to distinguish between $alpha_1$ (IV) and $alpha_2$ (IV) collagen polypeptide chains, a result that has not been visualized by gel permeation chromatography before (Fig.4).

A general image about the retention times of individual collagen species and some of their fragments is summarized in Table 1. The retention times decrease with increasing relative molecular mass, however, the decrease is not strictly linear in the logarithmic scale as it would be expected (Fig.5). No separation of collagen polypeptide chains of identical relative molecular mass but originating from different collagen types was observed. It has been communicated before that sorption plays an important role in high performance gel permeation chromatography. Indeed in our previous papers we have observed the separation of, e.g., alpha$_1$(I) and alpha$_1$(III) on the same sorbent that has been used in the present experiments when isocratic elution with 0.05 M Tris-HCl buffer, pH 7.5 (2M with respect to urea) was used. This was ascribed either to differences in hydrodynamic volumes of matching collagen peptide chains originating from different species or adsorption and/or partition interactions of the separated protein molecules, or finally to the possible weak affinity of collagen chains to the glucose coated macroporous adsorbent causing respective retention differences of otherwise similar molecules.

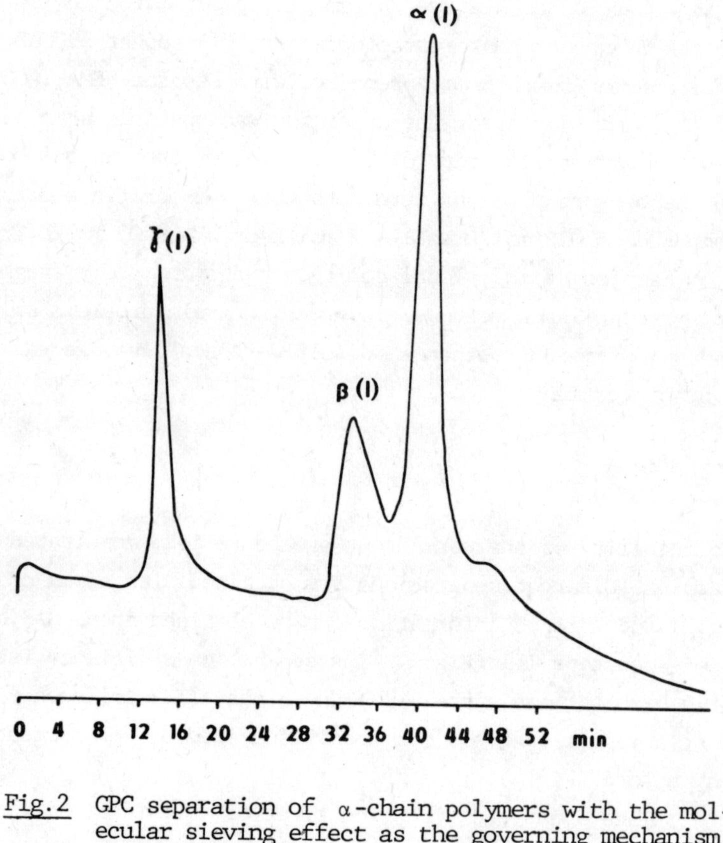

Fig.2 GPC separation of α-chain polymers with the molecular sieving effect as the governing mechanism of separation.

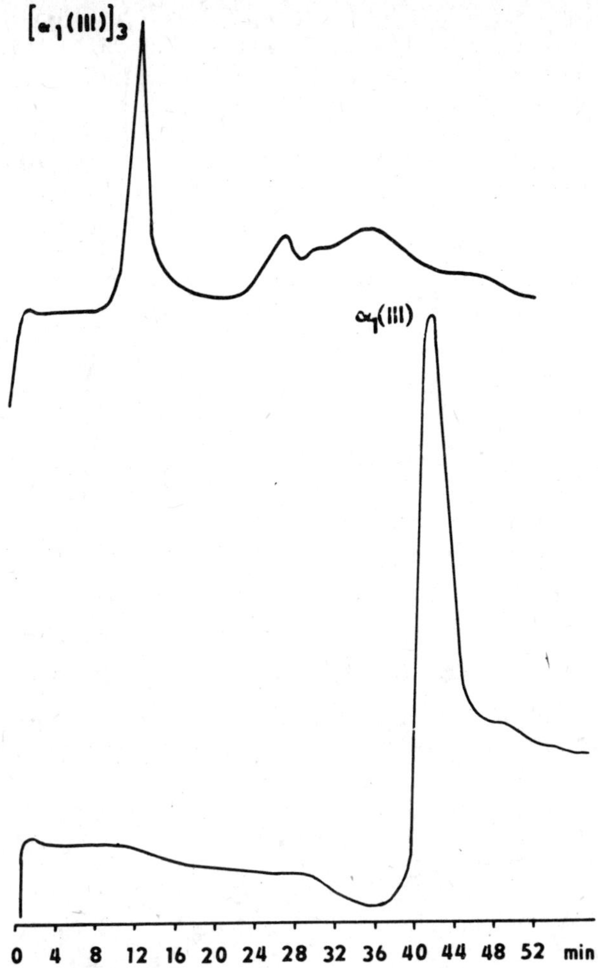

Fig.3 GPC of collagen type III in its trimeric (upper panel) and monomeric form (lower panel, after S-S bond cleavage)

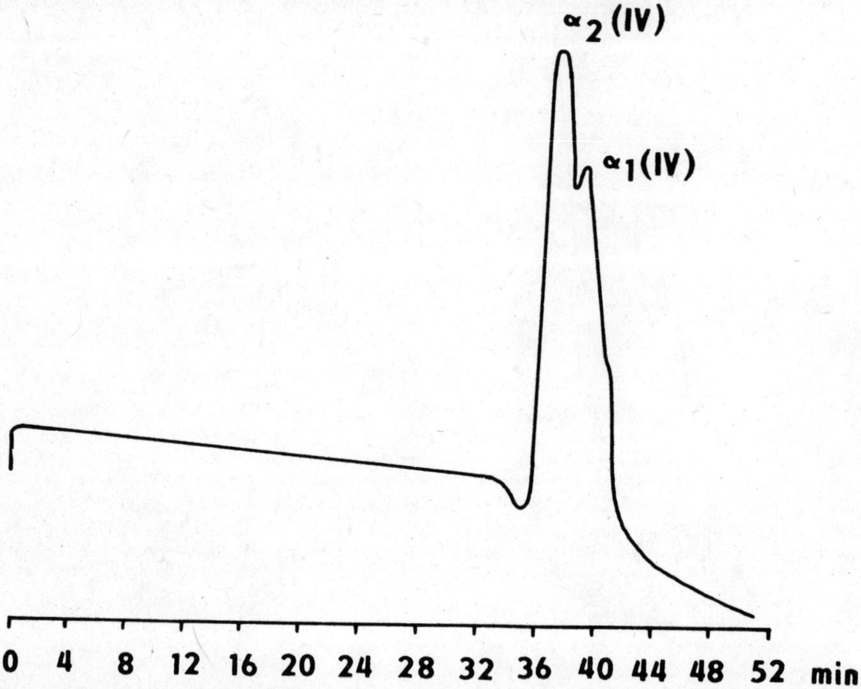

Fig.4 Separation efficiency of the HPGPC procedure demonstrated by the partial separation of both collagen type IV α–chains

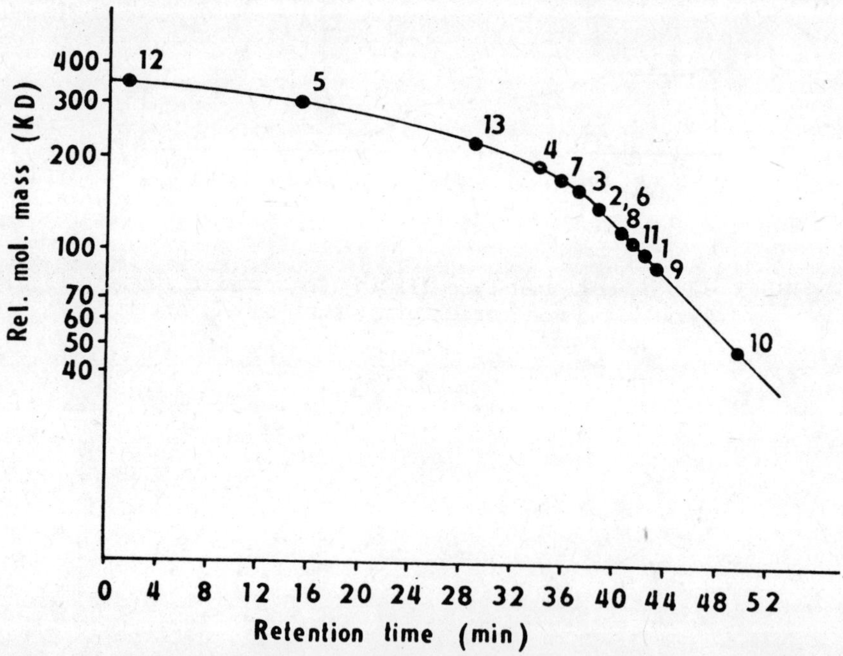

Fig.5 Log molecular mass vs. retention time plot of the collagen chains showing non-linearity of the relation

Table I Retention times of different polypeptide chains of the collagen family

	Type of collagen chain	Rel. mol. mass	Retention time min.
1	α_1(I), α_2(II), α_1(III)	100 000	42.0
2	α_1(IV) } N	140 000	38.5
3	α_2(IV) } N	160 000	37.0
4	β(I)	200 000	34.0
5	γ(I), $[\alpha_1(III)]_3$	300 000	14.0
6	α_1(IV) BM	160 000	37.0
7	α_2(IV) BM	180 000	34.5
8	C_1 fragment	120 000 av.	41.0
9	C fragment	95 000	43.0
10	50 K fragment	50 000	50.0
11	α_{1-3}(V)	110 000	41.5
12	7 S	360 000	4.0
13	7 S coll	225 000	29.0

It is evidenced now(Fig.6) that an increase of ionic strength of the eluent is capable of completely abolishing the inter-species differences at least between collagen I, III, IV and V. The same result can be achieved either by adding NaCl to the eluting solvent or by increasing the concentration of the Tris buffer to 0.5 Mol/L. The separation conditions can be selected in such a way that gel permeation is the only mechanism governing the separation. Still the non-linearity of the retention time vs. logarithm of molecular mass relation indicates that some other effects though minimalized still persist throughout the separation effected on Separon MEHA 1000 Glc.

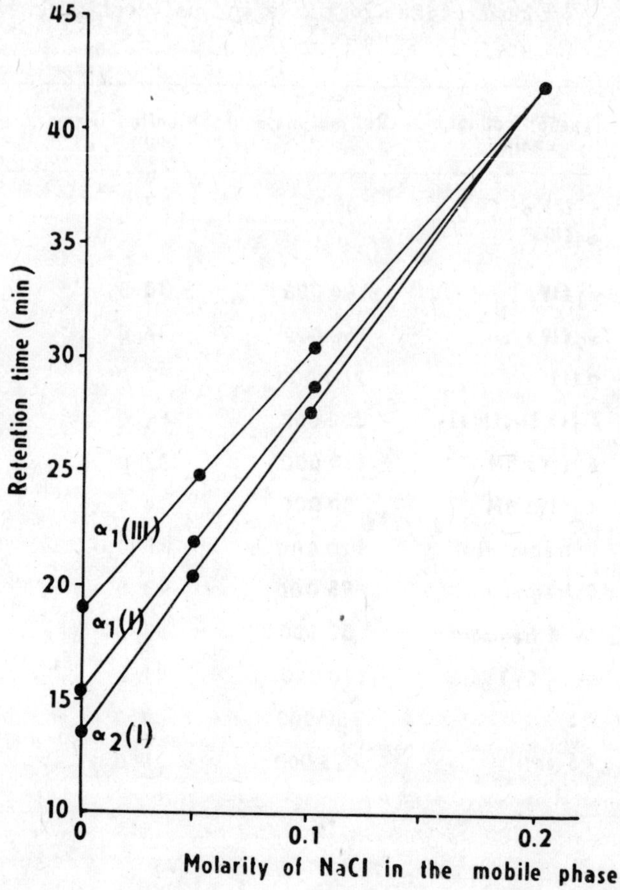

Fig.6 Dependency of the retention time upon the composition of the mobile phase

In other words with this category of gels it is possible to make a rational selection of conditions that would offer separations either solely on the basis of molecular sieving or may exploit other interactions as well offering thus separations of protein species of identical relative molecular mass. In our opinion this widens the separation possibilities not only with collagen proteins but with other protein mixtures as well.

REFERENCES

(1) FALON,A., LEWIS,R.V. and GIBSON,K.D.: Anal.Biochem. 110 318 (1981)

(2) MACEK,K., DEYL,Z., COUPEK,J. and SANITRAK,J.: J.Chromatogr.,222 284 (1981)

(3) GLANVILLE,R.W. in Conn. Tiss.Res., Chemistry, Biology and Physiology, Z.Deyl and M. Adam Eds, Alan R. Liss Inc. New York 1981, gp.1-14.

(4) DEYL,Z., HORAKOVA,M. and ADAM,M. in Conn.Tiss.Res.,Chemistry, Biology and Physiology, Z.Deyl and M. Adam Eds, Alan R.Liss Inc.New York 1981, gp. 15-44.

Proc. of the Symp. on Advances in Liquid Chromatogr., Szeged, Hungary, 1982

ADSORPTION CHROMATOGRAPHY OF FLEXIBLE POLYMER MOLECULES

G. GLÖCKNER

Department of Chemistry, Technical University of Dresden, Dresden, GDR

SUMMARY

TLC investigation of various polymers indicates reversible adsorption. The non-linearity of polymer adsorption isotherms is due to a cooperative effect. In column AC, the elution volume of poly(styrene-co-acrylonitrile) samples was found to vary according to the acrylonitrile content.

INTRODUCTION

As early as 1936, MARK and SAITO (1) filtrated solutions of cellulose acetate in acetone through columns packed with charcoal and isolated fractions from disintegrated zones of the packing by extracting with dioxane. But only a few papers dealing with the adsorption chromatography (AC) of polymers were published during a period of about 30 years after this inception. In 1968, BELENKIJ and GANKINA (2) as well as INAGAKI (3) independently performed thin-layer chromatographic separations of copolymers. Their publications stimulated a great number of successful TLC investigations of various problems concerning polymer size, composition, and architecture. During the last decade, numerous excellent baseline separations of oligomers were achieved by means of high performance column chromatography using either reversed phases (4-13) or bare silica (14-21) as a packing material. Separations of this kind were achieved for low-molecular poly-styrene (4, 5, 6, 14, 15, 16) and poly(methyl methacrylate) (17), for poly(ethylene oxide) (7, 8, 9), poly-(ethylene terephthalate) (18, 19), polyamide 6 (20), for epoxy resins (10, 11, 12), and polyesters (13). KLEIN and LEIDIGKEIT (21) even split up a mixture of some polystyrene (PS) samples (with molar mass values ranging between 2000 and 2000,000 g/mole) by means of AC.

These results are really remarkable because the adsorption isotherms of polymers are known to be extremely non-linear. They exhibit a pseudoplateau region which virtually covers the whole concentration range of interest. Calculations performed by FLEER and SCHEUTJENS (22) demonstrate that the pseudoplateau extends to very low concentrations. Its transition into the Henry region (with the amount adsorbed proportional to the concentration of the solution) occurs at a concentration which is the smaller the higher the molar mass of the polymer. For chains of only a hundred segments, the critical volume fraction is even as small as 10^{-25}. This means that solutions which contain more than one chain molecule in about 100 l solvent exhibit an adsorption which is nearly insensitive to a change in concentration. At a very low concentration, the adsorption constant is extremely large, but in the normal range of concentrations its differential value is not far from zero. As only a finite value of the distribution constant which does not vary too much with concentration enables the chromatographic process to be performed in a common manner, the AC of polymers will certainly not work under conditions employed in static adsorption measurements. Only under carefully selected conditions can reasonable values of polymer retention be reached. The adsorption energy of the solute must be balanced with the adsorption energy of the solvent, and the solvent/solute interaction energy.

INVESTIGATION INTO THE BEHAVIOUR OF MACROMOLECULES IN ADSORPTION CHROMATOGRAPHY

The behaviour of low-molecular solutes in AC can be described using the equation given by SNYDER (23):

$$\log K^{+} = \log V_a + \alpha_A(S^o - A_S \cdot \varepsilon^o) \qquad (1)$$

K^{+}: adsorption constant of the solute

V_a: volume of a monomolecular layer of the solvent on the surface of the adsorbent ("surface volume")

α_A: activity of the adsorbent $\alpha_A \leq 1$

S^o :adsorption energy of the solute

A_S: surface area covered by a solute molecule

ε^o: adsorption energy of the solvent

Among the prerequisites for this formula is the need for the reversibility of the adsorption process. If it can be proved that the SNYDER equation also holds true for polymers this would indicate that macromolecules are reversibly fixed under the conditions of AC. In order to perform this test, we looked for circumstances capable of balancing K^+ against V_a. The surface volume is about 10^{-2} cm^3/g. Consequently, the adsorption coefficient must also be small. In TLC measurements, this calls for high R_f values. We choose $R_f = 0.7$, which is certainly still on the safe side as concerns the volume profile on the plate. In TLC with pure solvents, polymers either remain at the starting spot or run with the solvent front. Binary mixtures of a strong solvent ($R_f = 1$) and a weak one ($R_f = 0$) also exhibit the one or the other extreme behaviour in nearly the whole range of possible compositions. The transition from $R_f = 0$ to $R_f = 1$ takes place in a rather narrow range. We investigated 6 polymers in 3 combinations of strong and weak solvents and obtained the results shown in Fig. 1 (24). The abscissa indicates the ε_M value of the mixture as calculated from the composition and ε^o values of the pure components. The ε_M which corresponded to $R_f = 0.7$ were evaluated by interpolation from these curves. These experimental data were averaged separately for each individual polymer and plotted vs. S^o/A_S. The latter were calculated for the respective polymer by the increments given by SNYDER (23). The straightforward plotting scheme follows from Eq. (1) if $\log K^+ = \log V_a$:

$$0 = \alpha_A(S^o - A_S \cdot \varepsilon^o) \qquad (2)$$

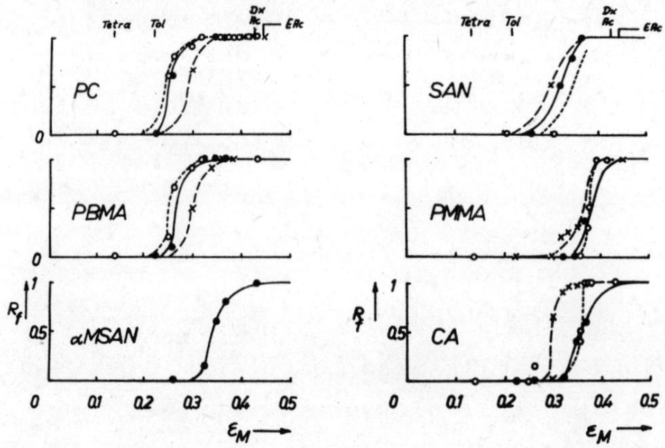

Fig.1

TLC results from 6 polymers on silica using the binaries tetrachloromethane/dioxane (o) toluene/acetone (●), and toluene/ethylacetate (x). PC: polycarbonate; PBMA: poly(butyl methacrylate); α MSAN: poly (α methyl-styrene-co--acrylonitrile), about 46 mole % AN; SAN: poly--(styrene-co-acrylonitrile), about 38 mole % AN; PMMA: poly(methyl methacrylate); CA: cellulose triacetate (24).

In this case, the activity α_A of the adsorbent does not matter, and the S^o/A_S values can be calculated for an arbitrary part of the polymer chain, for instance simply for one repeating unit. The straight-line in Fig. 2 indicates the equality of S^o/A_S and ε_M, following from the SNYDER equation under the condition given in Eq. (2).

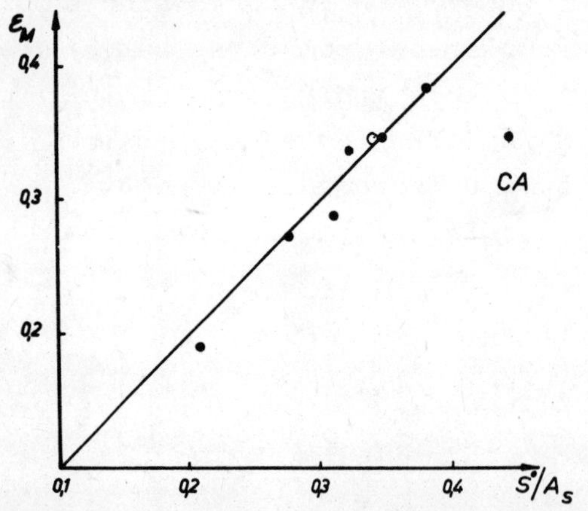

Fig. 2

Evaluation of results from Fig. 1 according to Eq. (2). The points correspond to the polymers PS, PC, PBMA, α MSAN, SAN, PMMA, CA (from left to right) (29)

The points obviously agree with this line, thus confirming the SNYDER formula as well as the reversibility of the adsorption. Modern theories of polymer adsorption also state reversibility, though the macromolecules cannot be removed from the surface by dilution. What is reversible is the adsorption process of any single segment. Linked to its neighbours, it cannot escape from the surface and will more likely be readsorbed than being so close to the surface until the other segments of the adsorbed train are eventually also desorbed. It is a kind of cooperative effect which makes a macromolecule unable to leave the adsorption layer. On the other hand, if macromolecules are capable of migration in AC technique they will certainly not touch the surface with a great many segments. In order to investigate this, we re-arranged Eq. (1).:

$$\ln K = -\frac{\Delta F}{kT} = 2.3 \log (K^*/V_a) = 2.3\,\alpha_A(S^o - A_S \cdot \varepsilon^o) \quad (3)$$

The derivative reads:

$$\frac{d \ln K}{d\,\varepsilon^o} = -\,2.3\,\alpha_A \cdot A_S \quad (4)$$

Figure 3 represents experimental results which BELENKIJ et al. (25) have obtained investigating polystyrene standards by means of TLC on silica with developing mixtures of cyclohexane, benzene, and acetone. The content of the strong solvent acetone was varied from 1.5 to 2.8 parts in mixtures with 56 part of the weaker components. At high values of ε_M, there is no adsorption, and the migration of the various PS probes is governed by an exclusion mechanism. The AC takes place in the region below the intersection point at $\varepsilon_M = 0.246$. Here, the experimental results can be represented by straight lines, the slope of which gives the derivative expressed by Eq. (4). (For numerical evaluation, the proportionality $\varepsilon^o_{SiO_2} = 0.77\,\varepsilon^o_{Al_2O_3}$ must be taken into account.)

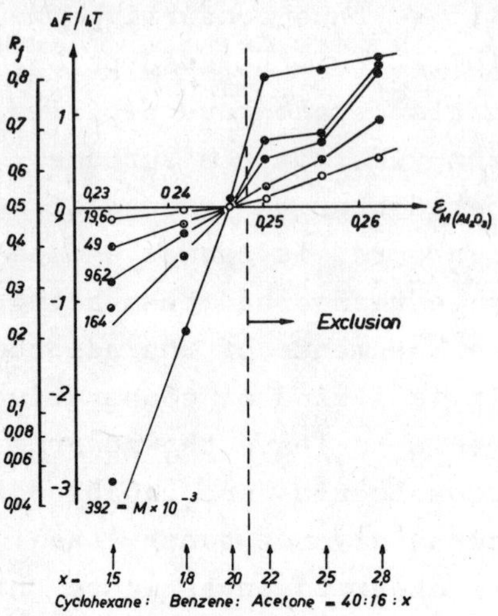

Fig.3.

TLC results obtained with 5 PS standards using mixed eluents on silica. Plot of $-\ln K = \Delta F/kT$ vs. $\varepsilon_M(Al_2O_3)$, according to BELENKIJ et al. (25)

These data and the molar mass of the samples are compiled in Table 1. For a first approximation we took $\alpha_A = 1$. Thus the data in column 2 can be understood as the surface area covered by the segments of one macromolecule. For one styrene unit, A_S amounts to 6.7. The figures in the third column indicate how often 6.7 is contained in the A_S value of the migrating PS sample.

Table 1

Slope of the straight lines in Fig. 3, and conclusions

M g/mole	$\frac{d(\Delta F/kT)}{2.3d\varepsilon^o_{SiO_2}}$	contacting styrene units per chain	styrene units per contact
19600	6.35	1	200
49000	19.1	3	165
96200	35.6	5	175
164000	57.6	9	185
392000	158.2	24	160

From this rough approximation, it can be concluded that macromolecules are absorbed with no more than 1 % of their repeating units under the conditions of AC.

SEPARATION OF COPOLYMERS ACCORDING TO CHEMICAL COMPOSITION

Separation according to molar mass will certainly not become the main task for polymer AC, because here there are effective competitive methods, as for instance SEC. The chemical composition distribution (CCD) of copolymers has been an important aim of many TLC investigations. By column AC on silica, TERAMACHI et al. (26) fractionated copolymers of styrene and methyl acrylate (MA). They found retention increasing with MA content using a linear gradient of methyl acetate (7 to 35 vol.%) in CCl_4. Employing internal standards, the authors were able to convert the elution curve into the curve of CCD, but in repeated experiments all three copolymer samples (46.6; 57.3; 77.9 mole % MA) were mostly eluted with a lower content of methyl acetate than in the preceding runs. DANIELEWICZ and KUBIN (27) investigated poly(styrene-co-methacrylate) samples. They employed a gradient of tetrahydrofuran (THF, 3 to 20 %) in dichloro ethylene. After each run, they flushed the silica in the column by at least 10 column volumes of pure THF and obtained reproducible retention values. Block copolymers of styrene and methyl methacrylate were investigated by BELENKIJ (28). The absorbent was silica, the solvent dichloro methane with increasing amounts of methanol added.

We investigated the behaviour of poly(styrene-co-acrylonitrile) in a silica column with THF in hexane.

MATERIAL AND METHODS

Copolymers: The specimens were polymerized in bulk with azo bisisobutyronitrile as an initiator. The polymerizations were performed in sealed ampoules which were shaken in a water bath of 60 °C. After conversion up to 5 - 10 %, the reaction was stopped by cooling the ampoules and puring the reaction mixture into stirred methanol. All the polymers re-precipitated twice and dried under vacuum at 60 °C. Table 2 gives data of the specimens used.

Table 2

Styrene/acrylonitrile copolymers investigated

sample	acrylonitrile content % per mole	M_{osm} g/mole
D68	7.9	244000
D47	27.4	325000
D55	37.8	297000
D38	59.4	340000

Solvents: Tetrahydrofuran, analytically pure, was stored with KOH for 24 hours, refluxed with Na for 2 hours and distilled. n-Hexane, a.p., as supplied by E. MERCK, Darmstadt.

For use as eluents, both solvents were mixed with 0.5 % methanol.

HPLC apparatus: HP 1080A (Hewlett Packard) with UV detector (254 nm). Column: 0,25 m long, 4.6 mm I.D., with LiChrospher$^{(R)}$ SI 100, d_p = 10 µm. Flow rate: 2 ml/min. Gradient elution with THF (20 to 90 %) in hexane.

Sample injection: 50 µl of solutions containing 0.4 g/l polymer in THF.

RESULTS

The chromatograms usually started with two sharp peaks, the first of which was due to some unretained polymer, the other to the injecting solvent. The elution volume of the retained polymer varied according to its acrylonitrile content. Fig. 4 gives some examples. As hexane is a non-solvent for the copolymers investigated, the runs began with the precipitation of the polymer. This happened without building up an additional flow resistance. The pressure change during the runs only represented the increase in viscosity due to the transition from 20 to 90 % THF. There was no influence from the polymer injected.

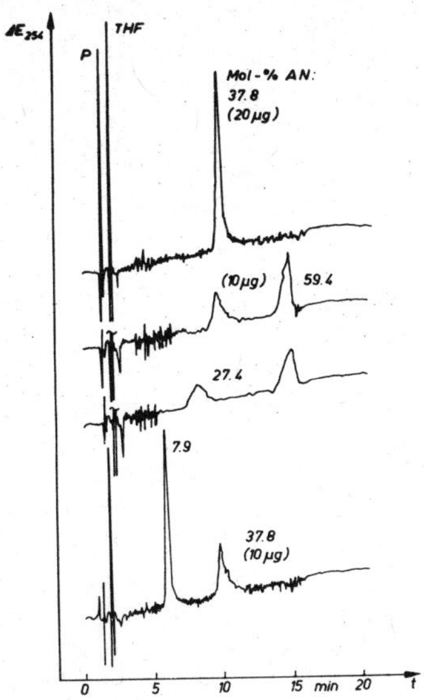

Fig. 4

UV record at 254 nm after injection of 20 µg SAN copolymer.
Upmost trace: aceotropic copolymer with 37.8 % AN, others: mixtures of two samples each, AN content indicated

The delivery of a certain polymer specimen from the column always took place at a certain THF content which corresponded to the THF content in a precipitating mixture. From this it can be concluded that the mechanism is solubility determined

DISCUSSION

Adsorbed polymers cannot be removed from the surface by dilution, though the adsorption is reversible for any segment. The desorption of macromolecules will easily advance if a competing substance with a higher level of adsorption energy is added. This substance will take over every adsorption site as soon as it is left by the polymer segment for a moment. So taking advantage of the dynamics of adsorption and desorption steps, the stronger solvent or polymer displaces the species adsorbed. From this it can be deduced that gradient elution should be the preferred mode of polymer adsorption chromatography. Indeed, this is clearly reflected by the literature.

In some TLC work, no effort was taken to produce any gradient, but this need not mean that the separation proceeded isocratically. In TLC with mixed eluents there are several possibilities of establishing a spontaneous gradient: via the vapour phase, by demixing during the run due to different velocity characteristics, or according to a difference in eluotropic strength. Such a hidden gradient can be rather meaningful in polymer studies. In the literature there are reports on the failure of isocratic column AC runs, which were seemingly performed under the same conditions as successful TLC separations.

Truly isocratic adsorption chromatography of polymers can only work on condition of a nearly complete balance between adsorption forces and solving power. Fine examples of such difficult work are the oligomer separations reported by EISENBEISS et al. (15) and by KNOX and McLENNAN (16).

REFERENCES

(1) MARK, H.; SAITO, G. (1936) Monatsch. Chemie 68 237-243

(2) BELENKIJ, B.G., GANKINA, E.S. (1969) Doklady Akademii Nauk SSSR 186 857

(3) INAGAKI, H., MATSUDA, H., KAMIYAMA, F (1968) Macromolecules 1 520-525

(4) KIRKLAND, J.J. (1975) Chromatographia 8 661-668

(5) PARRIS, N.A. (1978) 157 161-170

(6) LATTIMER, R.P., HARMON, D.J., WELCH, K.R. (1979) Analyt. Chem. 51, 1293-1296

(7) MELANDER, W.R., NAHUM, A., HORVATH, Cs. (1979) J. Chromatogr. 185 129-152

(8) NOZAWA, A., OHNUMA, T. (1980) J. Chromatogr. 187 261-263

(9) MURPHY, R., SELDEN, A.C., FISHER, M., FAGAN, E.A., CHADWICK, V.S. (1981) J. Chromatogr. 211 160-165

(10) SHIONO, S., KARINO, I., ISHIMURA, A., ENOMOTO, J. (198o) J. Chromatogr. 193 243-253

(11) HAGNAUER, G.L., SETTON, I. (1978) J. Liquid Chromatogr. 1, 55-73

(12) HAGNAUER, G.L. (1980) Polymer Composites 1 81-87

(13) COULOMBE, S., SCHAUWECKER, P., MARCHESSAULT, R.H., HOUTECOEUR, B. (1978) Macromolecules 11 279-281

(14) BECK, W., HALASZ, I. and FRESENIUS Z. (1978) Anal. Chem. 291 312-318

(15) EISENBEISS, F., DUMONT, E., HENKE, H. (1978) Angew. makromol. Chemie 71 67-89

(16) KNOX, J.H., McLENNAN, F. (1979) J. Chromatogr. 185 289-304

(17) ANDREWS, G.D., VATVARS, A. (1981) Macromolecules 14 1603--1605

(18) ZABORSKY, L.M. (1977) Analyt. Chem. 49 1166-1168

(19) HUDGINS, W.R., THEURER, K., MARIANI, T. (1978) J. Appl. Polymer Sci. 34 145-155

(20) BRODILOVÁ, J., ROTSCHOVÁ, J., POSPÍŠIL, J. (1979) J. Chromatogr. 168 530-532
(21) KLEIN, J., LEIDIGKEIT, G. (1979) Makromol. Chemie 180 2753-2756
(22) FLEER, G.J., SCHEUTJENS, J.M.H.M. "Adsorption of interacting oligomers and polymers at an interface" Lecture at the Canberra conference, April 1981, to be published in J. Colloid and Interface Sci., 1982
(23) SNYDER, L.R. (1968) Principles of Adsorption Chromatography Marcel Dekker, New York
(24) GLÖCKNER, G., MEISSNER, R. (1980) Acta Polymerica 31 191-192
(25) BELENKIJ, B.G., GANKINA, E.S., TANNIKOV, M.B., VILENČIK, L.Z. (1976) Doklady Akademii Nauk SSR 231 1147-1149
(26) TERAMACHI, S., HASEGAWA, A., SHIMA, Y., AKATSUKA, M., NAKAJIMA, M. (1979) Macromolecules 12 992-996
(27) DANIELEWICZ, M., KUBIN, M. (1981) 951-956
(28) BELENKIJ, B.G. (1979) Pure and Appl. Chem. 51 1519-1535
(29) GLÖCKNER, G. (1980) J. Polymer Sci. C, Polymer Symposia 68 179-183

Proc. of the Symp. on Advances in Liquid Chromatogr., Szeged, Hungary, 1982

THE EVALUATION OF MOLECULAR CONNECTIVITY INDICES AND VAN DER WAALS VOLUMES FOR CORRELATION OF CHROMATOGRAPHIC PARAMETERS

J. BOJARSKI, L. EKIERT

Department of Organic Chemistry,
Nicolaus Copernicus School of Medicine,
Kraków, Poland

The size and shape of molecules are considered as important factors in the molecular interactions involved both in structure-activity relationships in medicinal chemistry and in the behaviour of molecules during chromatographic separation processes.

One of the parameters successfully used in QSAR studies is the molecular connectivity index (1,2)- a topological parameter that quantitates various structural features, such as: size, branching, cyclization, unsaturation and heteroatom content of the molecule. There are different types and orders of this parameter widely used for the correlations and predictions of chromatographic data (3-5).

Another parameter describing the molecule is the Van der Waals volume (6) also used in QSAR studies (7,8).

These parameters are represented by the expressions:

$$^{1}\chi^{V} = \Sigma_{s=1}^{N} (\delta i \, \delta j)^{-\frac{1}{2}}$$

$$V_W = \Sigma V_W^A + \Sigma C$$

where $^{1}\chi^{v}$ is the first order valence molecular connectivity index, N is the total number of bonds in the molecule, δ is a value corresponding to the difference between the number of valence electrons and the number of hydrogen atoms attached to i-th and j-th atoms, respectively, V_W is the Van der Waals volume of the molecule, V_W^A is the spherical volume of the atom calculated from its Van der Waals radius and C is the correction of Van der Waals volume of the atom for sphere overlapping due to covalent bonding and for branching (7).

The relation of molecular connectivity to molecular volume and biological activity was recently studied by Hall and Kier (9).

We wanted to study the application of these molecular parameters and their predictive value for the correlations with chromatographic data of barbituric acid derivatives represented by the formula:

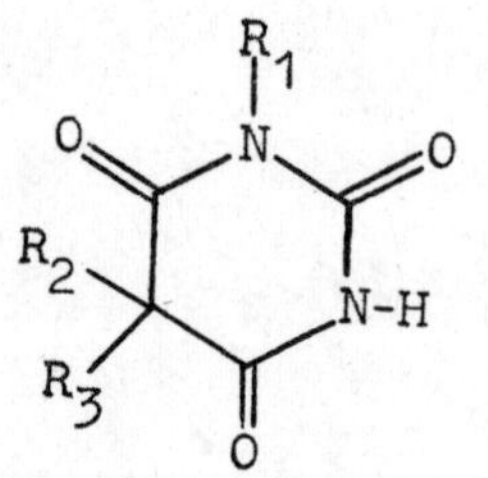

R_1 = H, CH_3

R_2, R_3 = aliphatic and cyclic hydrocarbon substituents

The data for these studies were taken from the literature and a part of the results dealing only with the connectivity indices was recently published (10). The studies reported here include the correlations of chromatographic data with the Van der Waals volumes and the comparisons between these parameters.

The following numbers of data sets and the following chromatographic parameters were used for the correlations:

paper chromatography	- 18	- R_F and R_M
thin-layer chromatography	- 12	- R_F and R_M
gas chromatography	- 52	- I, t_R and log t_R
high-performance liquid chromatography	- 24	- k; log k; t_R and log t_R

Barbituric acid derivatives were subdivided into 4 groups as follows:

group A - aliphatic saturated substituents at C_5

group B - A + aliphatic unsaturated substituents at C_5

group C - B + N-methyl C_5 disubstituted compounds with aliphatic and cyclic /including aromatic/ substituents

group D - C + cyclic substituents at C_5

The correlations were carried out by the least squares method according to the equations:

$$y = a \, {}^{1}\chi^{v} + b$$

$$y = a \, V_W + b$$

where y is the chromatographic parameter, ${}^{1}\chi^{v}$ is the first order valence connectivity index, V_W is the Van der Waals volume of the molecule and a and b are the slope and the intercept of the straight line, respectively.

The results of correlations are exemplified by those for HPLC data presented in Table 1.

Table 1
Correlation coefficients for HPLC chromatographic data of barbiturates

r	% of correlations							
	${}^{1}\chi^{v}$				V_W			
	A	B	C	D	A	B	C	D
> 0.99	42				21			
0.95 - 0.99	17	25	8		29	29	8	
0.90 - 0.94	41	29	29	4	42	25	33	8
0.80 - 0.89		33	38	13	4	25	50	42
0.70 - 0.79		13	25	50	4	13	4	42
0.60 - 0.69				25		4	4	4
0.50 - 0.59				8		4		4
95% confidence level	79	100	100	100	79	92	96	96

Table 2

Correlation coefficients for $^1\chi^v$ vs V_W parameters

r	PC				TLC				GC				HPLC			
	A	B	C	D	A	B	C	D	A	B	C	D	A	B	C	D
$V_W > {}^1\chi^v$	12	6	11	44	39	54	65	83	12	58	50	17	4	37	58	88
$^1\chi^v > V_W$	63	83	89	45	40	46	35	15	50	31	38	81	58	38	33	8
$^1\chi^v = V_W$	25	11	-	11	21	-	-	2	38	11	12	2	38	25	9	4

Table 3

Significance of correlation coefficients for $^1\chi^v$ vs V_W parameters

	PC				TLC				GC				HPLC			
	A	B	C	D	A	B	C	D	A	B	C	D	A	B	C	D
$V_W > {}^1\chi^v$	-	-	-	11	6	47	50	71	4	29	21	12	-	4	21	42
$^1\chi^v \quad V_W$	25	50	67	28	26	33	31	7	33	11	15	46	33	25	8	8
$^1\chi^v = V_W$	75	50	33	61	68	20	19	22	63	60	64	42	67	71	71	50

$>$ better than

The results for other chromatographic techniques, taking into account the significance of correlation coefficients, were less satisfactory and the worse were those for TLC. The dispersion of correlation coefficients for the subgroups of compounds was similar for all chromatographic techniques.

The comparisons of the utility of molecular connectivity indices and Van der Waals volumes for the correlations of chromatographic data of barbiturates were made in terms of correlation coefficients and their significance. The results are presented in Tables 2 and 3 as the percentage of correlations where the connectivity index yields better, equally good or worse results than the Van der Waals volume.

From Table 2 it can be seen that the correlation coefficients are better, the same or worse using the connectivity index or Van der Waals volume. depending on the chromatographic technique but also on the structural subgroups of barbiturates. No definite advantage of one parameter over the other can be deduced from these data.

The significance levels of correlation coefficients are generally better for the connectivity index for paper chromatography but worse for TLC for almost all subgroups. No equivocal results were obtained for GC and HPLC and quite often the significance levels were the same.

When we assume 95 % confidence level as a certain estimate of the quality of correlation coefficients and look on the percentage of these coefficients that are at or above that level for each parameter, it is evident that the majority of the data are correlated reasonably well using both parameters /Table 4/.

It follows from these results that both molecular connectivity indices and Van der Waals volumes may be used for correlations of chromatographic data when a good quality of these data is achieved.

One can search for better correlations purposely changing various parameters of chromatographic separation processes and trying to find those essential for that "good quality". Another approach implies multiparameter correlations using higher order connectivity indices, or an introduction of other parameters describing the chromatographic process or simply a change in the rules of calculation of the topological index. This last case may be illustrated by the work of Stead et al. (11). They tried to correlate gas chromatographic retention data for 5,5-disubstituted barbiturates using the standard first order valence connectivity index and found good correlations in the subgroups with similar substituents. The whole set of data, hovewer, was much better correlated by the modified first order valence connectivity parameter $^{1}\chi^{v}_{N}$. The calculation rules of that parameter were identical as those for the classical first order valence connectivity index for saturated aliphatic substituents but different treatment was proposed for unsaturated and cyclic substituents.

Table 4

Percentage of correlations with r at 95 % confidence level for $^{1}\ ^{v}$ and/or V_W

α 95 %	PC				TLC				GC				HPLC			
	A	B	C	D	A	B	C	D	A	B	C	D	A	B	C	D
V_W	-	-	-	-	-	14	7	35	2	2	7	3	-	-	-	-
χ	12	6	11	6	4	-	-	-	4	-	3	8	4	8	4	4
χ, V_W	88	94	89	94	96	86	93	65	94	98	90	89	96	92	96	96

We studied the application of this modified connectivity index for correlations of 25 sets of chromatographic data of barbiturates and compared the results with those obtained by us for classical $^{1}\chi^{v}$ values. The results are shown in Table 5 as the number of correlations better for each molecular connectivity parameter.

Table 5

Number of correlations for $^{1}\chi^{v}$ vs $^{1}\chi^{v}_{N}$ parameter

	PC	TLC	GC	HPLC
$^{1}\chi^{v}$ better than $^{1}\chi^{v}_{N}$	6	8	-	6
$^{1}\chi^{v}_{N}$ better than $^{1}\chi^{v}$	-	-	5	-

From these results it follows that the modified valence connectivity index $^{1}\chi^{v}_{N}$ yields better correlations only for the GC technique, while the liquid chromatography data are better correlated by the classical valence connectivity indices.

One may guess that the advantage of one connectivity index over the other for these chromatographic techniques should be a reflection of differences between the separation processes when gas or liquid phases are involved, but further research is needed to validate this hypothesis. In conclusion, the molecular connectivity indices and Van der Waals volumes may be useful in describing the interactions between member molecules of one family and the stationary phases, but the validity of such correlations depends on the quality of chromatographic data and on the structural parameters employed.

REFERENCES

(1) KIER,L.B., HALL,L.H. (1976) "Molecular Connectivity in Chemistry and Drug Research", Academic Press, New York, N.Y.

(2) KIER,L.B., HALL,L.H. (1981) J. Pharm.Sci., 70 583 and references cited therein.

(3) BUYDENS,L., MASSART,D.L. (1981) Anal.Chem. 53 1990 and references cited therein.

(4) WELLS,M.J., CLARK,C.R.,PATTERSON,R.M.(1982) J.Chromatogr.,235 61

(5) HURTUBISE,R.J.,ALLEN,T.W., SILVER,H.F.(1982) J.Chromatogr.235 517

(6) BONDI,A. (1964) J.Phys.Chem. 68 441

(7) MORIGUCHI,I.,KANADA,Y., KOMATSU,K. (1976) Chem.Pharm.Bull. 24 1799

(8) BINDAL,M.C., SINGH,P.,BHATNAGAR,R.P.,GUPTA,S.P. (1980) Arzneim. Forsch. 30 924

(9) HALL,L.H., KIER,L.B. (1981) Eur.J.Med.Chem. 16 399

(10) BOJARSKI,J., EKIERT,L. (1982) Chromatographia 15 172

(11) STEAD,A.H., GILL,R., EVANS,A.T., MOFFAT,A.C. (1982) J.Chromatogr., 234 277

DISPLACEMENT CHROMATOGRAPHY

Proc. of the Symp. on Advances in Liquid Chromatogr., Szeged, Hungary, 1982

EFFICIENCY OF SEPARATION PROCESSES AS APPLIED TO DISPLACEMENT CHROMATOGRAPHY

G.E. VERESS,*,** CS. HORVÁTH,** E. PUNGOR*

*Institute for General and Analytical Chemistry, Technical University, Budapest, Hungary
**Department of Chemical Engineering, Yale University, New Haven, Connecticut, USA

SUMMARY

In order to express the performance of simple separation processes two sets of efficiency parameters are introduced and their relationship is discussed and illustrated graphically. The treatment is applied to both binary and multicomponent separations involving single stage and multistage processes and it leads to a general comparison of the effectiveness of various separation systems. In particular, this approach is used to analyze the efficiency of separation by displacement chromatography when certain simplifying conditions hold. It is shown that with multicolumn systems the recycling mode of operation yields the highest rate of production.

INTRODUCTION

The recent renaissance of displacement chromatography stems from the possibility of carrying out preparative scale separations with columns and instrumentation employed in analytical HPLC (1,2). Unlike in conventional elution chromatography, the feed concentration is high in this technique and the mixture is not separated into bands of the individual components that move with different velocities through the column but into adjacent zones, all moving with the same velocity determined by that of the displacer front (3,4). Whereas the separation efficiency in elution chromatography has profusely been treated in the literature, no such treatment has been put forward for displacement chromatography.

In this report a general framework is given for the characterization of the effectiveness of separation processes. The results are applied to displacement chromatography to express its efficiency and to find the type of multicolumn arrangement that yields the highest rate of production.

EFFECTIVENESS OF SEPARATION PROCESSES

Purification

The schematics of a simple separation, purification process is depicted in Fig.1. The material balance is expressed by

$$x = x_p + x_\ell \qquad /1/$$

where x, x_p and x_ℓ are the amounts of feed, pure product and unresolved mixture, respectively, and the following inequalities apply

$$0 \leq x,$$
$$0 \leq x_p \leq x,$$
$$0 \leq x_\ell \leq x.$$

The effectiveness of the above separation process will be characterized by types of parameters. The first type takes into account both input and output. The effectiveness of the separation is expressed by the yield of pure product, e, as

$$e = x_p/x \qquad /2a/$$

whereas the ineffectiveness of the separation, i, is defined as

$$i = x_\ell \;/\; x \qquad /2b/$$

Thus

$$e = x/(x_p + x_\ell) = (x - x_\ell) \;/\; x \qquad /3a/$$

and

$$i = x_\ell \;/(x_p + x_\ell) = (x - x_p)/x \qquad /3b/$$

Material balance yields the relationship

$$e + i = 1 \qquad /4/$$

with the following inequalities

$$0 \leq e \leq 1,$$
$$0 \leq i \leq 1.$$

The other type of performance characteristics is based on the output only. The separation effectiveness ratio is defined as

$$\varepsilon = x_p \;/\; x_\ell \qquad /5a/$$

where the separation ineffectiveness ratio is expressed by

$$\omega = x_\ell \;/\; x_p \qquad /5b/$$

It follows from the material balance that

$$\varepsilon = x_p \;/\; (x - x_p) = (x - x_\ell)/\; x_\ell \qquad /6a/$$

and $$\omega = (x - x_p) \;/\; x_p = x_\ell \;/\; (x - x_\ell) \qquad /6b/$$

Combination of eq. 5a and 5b gives that

$$\varepsilon \cdot \omega = 1 \qquad /7/$$

and the pertinent inequalities are

$$0 \leq \varepsilon \leq +\infty$$

$$0 \leq \omega \leq +\infty$$

The effectiveness and ineffectinevess ratios can be expressed by the yield and the ineffectiveness of the process as

$$\varepsilon = e/i \qquad /8a/$$

and

$$\omega = i/e \qquad /8b/$$

A graphical illustration of the relationship between the above parameters is given in Fig.2. The efficiency of the process is represented by points in the effectiveness - ineffectiveness space on a straight line, i = 1-e, according to eq. 4 and the appropriate inequalities.

The "effectiveness line" can be scaled to become the locus of the parameters ε and ω, since at the point M(1/2,1/2) they are equal to one, at B/0,1/ $\varepsilon = 0$ and $\omega = +\infty$, whereas at G/1,0/ $\varepsilon = +\infty$ and $\omega = 0$.

Accordingly the points B, M and G represent poor, intermediate and complete separation, respectively. For any arbitrarily chosen point P/e.i/ the above parameters can be expressed as

$$\varepsilon = e/i = \mathrm{tg}\ \beta = \mathrm{ctg}\ \alpha \qquad /9a/$$

and

$$\omega = i/e = \mathrm{ctg}\,\beta = \mathrm{tg}\ \alpha \qquad /9b/$$

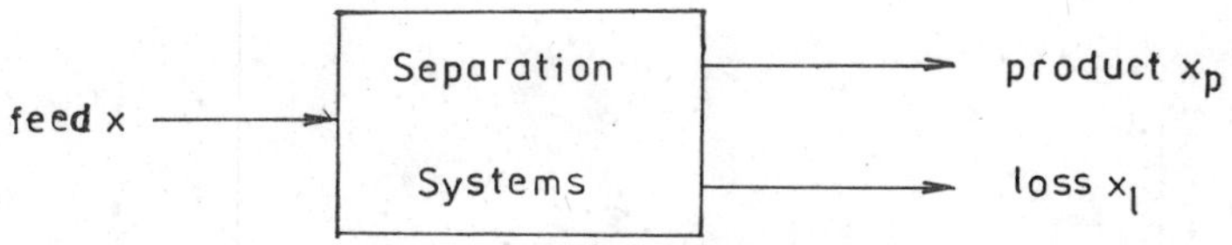

Fig.1. Schematics of a simple purification system

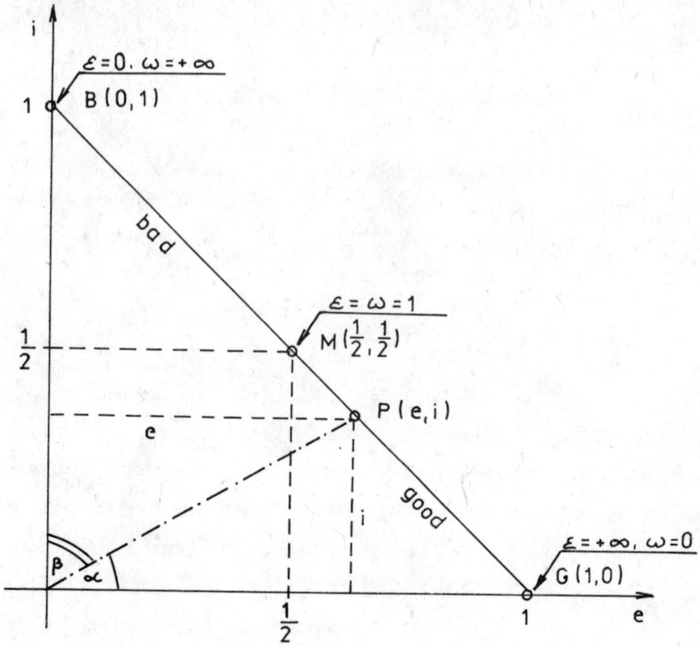

Fig.2. Graphical illustration of the relationship between the measures of separation efficiency

Multicomponent separations

The four measures of separation efficiency or inefficiency can be applied to multicomponent separations as well. They can be defined for any component j as follows:

$$e^j = x_p^j \;/\; x^j \qquad /10a/$$

$$i^j = x_\ell^j \;/\; x^j \qquad /10b/$$

$$\varepsilon^j = e^j \;/\; i^j \qquad /10c/$$

$$\omega^j = i^j \;/\; e^i \qquad /10d/$$

alternatively they can be written for the entire mixture containing s components as

$$e^{total} = \frac{\sum_{j=1}^{s} x_p^j}{\sum_{j=1}^{s} x_j} = x_p \;/\; x \qquad /11a/$$

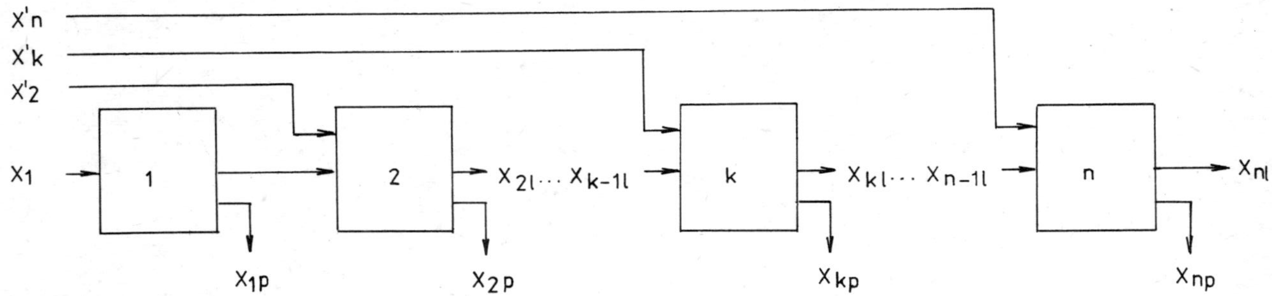

Fig.3. Schematics of a multistage separation process

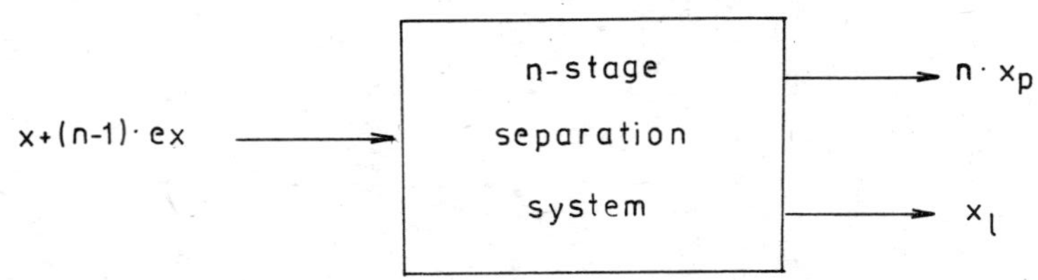

Fig.4. General representation of multistage separation systems

$$i^{total} = \frac{\sum_{j=1}^{s} x_{\ell}^{j}}{\sum_{j=1}^{s} x^{j}} = x_{\ell} \; / \; x \qquad /11b/$$

$$\varepsilon^{total} = e^{total}/i^{total} \qquad /11c/$$

$$\omega^{total} = i^{total}/\, e^{total} \qquad /11d/$$

Multistage systems

In order to extend the scope of the above treatment to multistage separation processes such as the recycling system shown in Fig.3. the following assumptions are made. The feeds are additive, i.e.,

$$x_k = x'_k + x_{k-1\ell} \qquad /12a/$$

the feeds are equal, i.e.,

$$x_1 = x_2 \ldots = x_k = \ldots x_n = x \qquad /12b/$$

and the stage effectiveness is invariant, i.e.,

$$e_1 = e_2 = \ldots e_k = \ldots = e_n = e \qquad /12c/$$

and that e is not unity.

From eq. 12a-12c it follows that

$$x_{1p} = x_{2p} = \ldots = x_{kp} = \ldots = x_{np} = x_p \qquad /13a/$$

$$x_{1\ell} = x_{2\ell} = \ldots = x_{k\,\ell} = \ldots x_{n\ell} = x_{\ell} \qquad /13b/$$

$$x'_2 = \ldots = x'_k \ldots = x'_n = x' \qquad /13c/$$

where

$$x_p = e, x \qquad /13d/$$

and

$$x' = x_p = e,\, x \qquad /13e/$$

A simplified schematic representation of a multistage /recycling/ separation system is shown in Fig.4. The pertinent performance measures ε /k/ and ω /k/ are given by

$$\varepsilon\,/k/ = k \cdot \varepsilon \qquad /14a/$$

and

$$\omega/k/ = \omega\ /k \qquad /14b/$$

where the stage number k goes from one to n.

Equs 14a and 14b lead to the following expressions for yield and separation ineffectiveness as

$$e\left[k\right] = \frac{1}{1+\omega\left[k\right]} = \frac{1}{1+\frac{\omega}{k}} \qquad /16a/$$

$$i\left(k\right) = \frac{1}{1+\varepsilon\left[k\right]} = \frac{1}{1+k\cdot\varepsilon} \qquad /16b/$$

The calculation of these parameters can also be carried out by the recursive relationships:

$$e\left[k\right] = \frac{1+\frac{k-e}{(k-1)^2}\cdot\frac{1}{e}}{1+\frac{k-1}{(k-1)^2}\cdot\frac{1}{e}} \cdot e\left[k-1\right] \qquad /17a/$$

$$i\left[k\right] = \frac{1+(k-2)\ i}{1+(k-1)\ i}\ i\left[k-1\right] \qquad /17b/$$

where k goes from 2 to n.

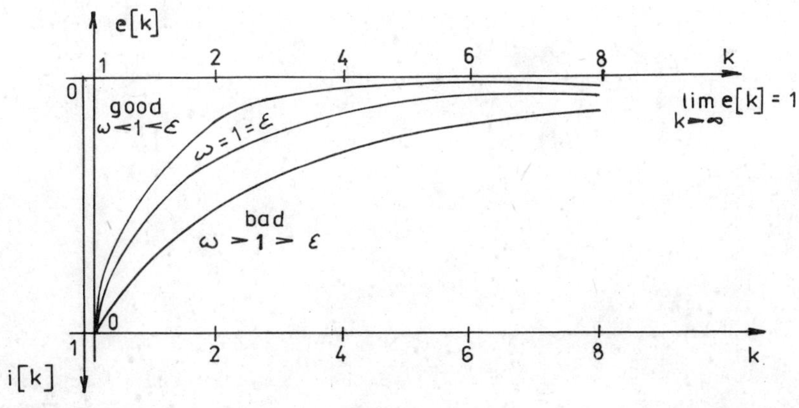

Fig.5. Effectiveness and ineffectiveness in multistage separations as a function of the number of stages

The dependence of the efficiency parameters on the number of stages is illustrated in Fig.5. As in the case of single stage processes discussed previously the effectiveness parameters for multistage separation systems can be defined either for a key component or for the total mixture.

DISPLACEMENT CHROMATOGRAPHY

Typical result of the separation process are illustrated by the chromatogram in Fig.6. The individual components emerge from the column as overlying bands of pure products in the carrier. Thus between each component j and j + 1 there is a mixed zone m^{jj+1} the amount of which is given by

$$m^{jj+1} = x^{jj+1} + x^{j+1j} \qquad /18a/$$

where x^{jj+1} and x^{j+1j} are the unresolved amounts of the above two components, respectively.

For component j the amount of recovered pure product is x_p' and the unresolved amount of this component is given by

$$x^{j} = x^{jj-1} + x^{jj+1} \qquad /18b/$$

where x^{jj-1} and x^{jj+1} are the tail and front of band j in the mixed zone, respectively, and can be evaluated from the chromatogram.

If these data are not available, i.e. the tail and front ends are not known for each zone, we may assume that they are equal to

$$x^{jj-1} + x^{jj+1} = x^{j-1j} + x^{j+1j} = x_\ell^j \qquad /19a/$$

It follows from equs 18a and 19a that

$$x^{j-1j} + x^{jj-1} + x^{jj+j\ell} + {}^{j+1j} = 2x_\ell^j \qquad /19b/$$

that is

$$m^{j-1j} + m^{jj+1} = 2x_\ell^j \qquad /19c/$$

or

$$x_\ell^j = \frac{1}{2}\left(m^{j-1j} + m^{jj+1} \right) \qquad /19d/$$

with the values of x_p^j and x_ℓ^j at hand the effectiveness of displacement chromatography can be measured by the parameters defined above. In many instances the yield e is of interest only for one or a few products from a multicomponent mixture. However, when the number of components is small the evaluation of e^{total} may be appropriate for characterizing the performance of the displacement chromatographic system.

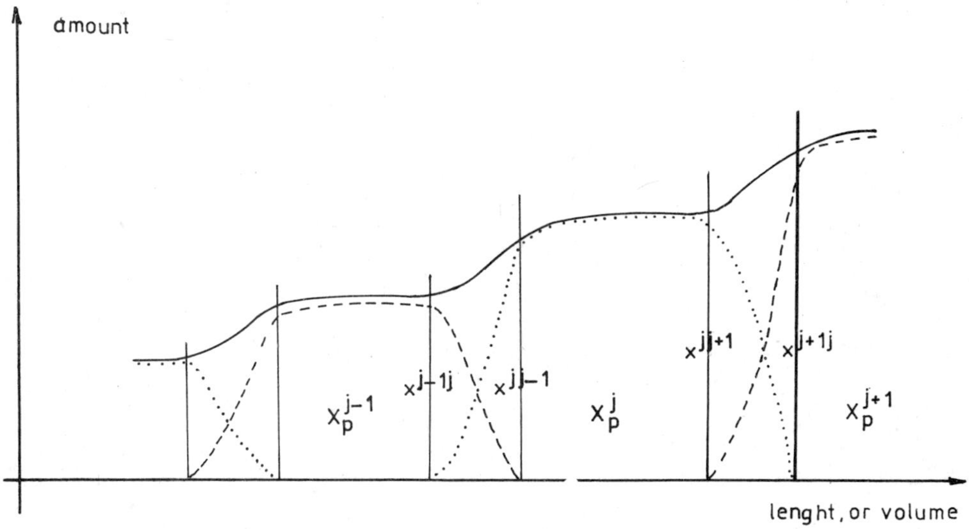

Fig.6. Typical chromatogram obtained in displacement chromatography. The areas of the component zones represent the amount of material

Multistage displacement chromatography in production scale displacement chromatography is preferably carried with more than one column in a multistage operation. In such an arrangement some of the columns are being used for the displacement development while the others are being regenerated. Thus in a system comprising n columns each of them will be utilized for the actual separation and regenerated for the time periods t_w and t_r, respectively.

Our interest is to find the operational scheme, which gives the maximum rate of production, among the following three arrangements:

Scheme A, n/2 columns separate and n/2 columns regenerate at the same time;

Scheme B, first n columns separate simultaneously then all of them regenerate at the same time;

Scheme C, a column cascade with recycling is employed.

Table I. Comparison of different displacement chromatographic systems comprising n columns.

Scheme	Working arrangement	Process time per run	Amount of pure product per run	Production rate
A	One half of columns separate while the other half regenerate	$\max\{t_W, t_R\}$	$\frac{n}{2} \cdot e \cdot x$	$\frac{n\, e\, x}{2 \max\{t_W, t_R\}}$
B	All columns separate and subsequently regenerate	$t_W + t_R$	$n \cdot e \cdot x$	$\frac{n \cdot e \cdot x}{t_W + t_R}$
C	Columns form a cascade with reaching	$t_W + t_R$	$n \cdot \frac{ne}{ne+1-e} \cdot x$	$\frac{n \cdot \frac{ne}{ne+1-e} \cdot x}{t_W + t_R}$

Characteristics of the above operational schemes are listed in Table 1. It is seen that scheme C offers the highest rate of production. With the exception of the situation, where $t_w = t_r$ and thus schemes A and B give the same production rate, scheme B is superior to scheme A as seen by comparing the rates of production shown in Table 1.

ACKNOWLEDGEMENTS

G.E. Veress was at the Yale University as a research fellow of the Mutual Educational Exchange Grant made under the provisions of the Fulbright Program, administered by the Council for International Exchange of Scholars.

This work was supported in part by Grant GM 20993 from the National Institute of Health, U.S. Public Health Service, Department of Health and Human Service.

NOTATION

B/0,1	point of poor separation in Fig.2
e	separation effectiveness or yield of pure product
G/1,0/	point of complete separation in Fig.2
e [k]	separation effectiveness of k-stage system
i	separation ineffectiveness
i [k]	separation ineffectiveness of k-stage system
m	mixture of two components in zone overlap
M/1/2,1/2/	point of intermediate separation in Fig.2
n	number of stages
P/e,i/	arbitrary point of separation in Fig.2
s	number of the components
t_r	regenerating time
t_w	working time
x	amount of feed or amount in general
x'	amount of additional feed to a stage
α	angle belonging to the separating point/see Fig.2/
β	complementary angle belonging to the separation point /see Fig.2/
ε	separation effectiveness ratio
ε [k]	separation effectiveness ratio of k-stage system
ω	separation ineffectiveness ratio
ω [k]	separation ineffectiveness ratio of k-stage system

SUBSCRIPTS

k	stage number
ℓ	unresolved fraction
p	product

SUPERSCRIPTS

j	component number
total	index for the whole mixture

REFERENCES

(1) HORVATH,Cs., NAHUM,A. and FRENZ,J.H. J.Chromatogr./1981/ 218, 365
(2) KALASZ,H. and HORVATH,Cs. J.Chromatogr. /1981/ 215, 295
(3) TISELIUS,A. and HAGDAHL,L. Acta Chem.Scand. /1950/ 3, 394
(4) HELFFERICH,F. and KLEIN,G., Multicomponent Chromatography-Theory of Interference, Marcel Dekker, New York, 1970, pp. 225-243.

Proc. of the Symp. on Advances in Liquid Chromatogr., Szeged, Hungary, 1982

EFFECT OF OPERATING CONDITIONS IN DISPLACEMENT THIN-LAYER CHROMATOGRAPHY

H. KALÁSZ*, CS. HORVÁTH**

*Department of Pharmacology, Semmelweis University of Medicine, Budapest, Hungary
**Department of Chemical Engineering, Yale University, New Haven, Connecticut, USA

SUMMARY

Operational steps in displacement development by thin-layer chromatography have been investigated. In most cases carrier displacement of certain phenylalkylamines was carried out by using the components of a dye-mixture as the carrier substances, chloroform as the carrier solvent and triethanolamine as the displacer. The results showed that without presaturation of the chamber with the vapor of the developer solvent the migration of the displacer front relative to that of the carrier solvent increases. Thus the R_D value can be increased not only by increasing the concentration of the displacer but also by carrying out the development without presaturation. The separation obtained with this technique was not influenced by the sample size up to 1.5 mg under the conditions investigated and satisfactory results were obtained with sample strips 3 cm in length. It was found that the height of the thin-layer plate can be much smaller, when the separation is performed in the displacement mode, than that required in the conventional elution mode as each step of displacement development takes a relatively small distance on the plate. Furthermore a concentration of the sample components can take place in the course of displacement development and therefore the technique offers a potential in both trace analysis and preparative separation. For displacement chromatography of phenylalkylamines different combinations of carrier solvent and displacer were examined. Whereas all ethanolamines investigated exhibited similar properties as displacers the addition of acetone to chloroform resulted in dramatic changes in the migration behaviour of the sample components.

INTRODUCTION

Ensconced as a popular and versatile analytical technique, thin-layer chromatography (TLC) has traditionally been carried out in the elution mode in order to resolve sample components into individual spots that migrate with different velocities on the plate (1). Only very recently it has been demonstrated that the scope of two-dimensional TLC can be extended by employing displacement development and the use of carrier displacement can lead to novel type of separations by TLC (2).

The displacement mode of chromatography had been established (3-5) before elution became the principal mode for the development of chromatogram. Inadequacies of the sorbents available, relative poor column efficiency and lack of suitable methods for detection, however, had thwarted displacement chromatography to become a practical separation tool. Only after high efficiency columns and precision instrumentation have been developed for HPLC did interest awake anew in displacament chromatography. A recent series of papers has demonstrated that high performance displacement chromatography is a powerful technique to perform preparative scale separations with columns and equipment employed in analytical HPLC (6-9). Furthermore, advances have been made in the study of the theory and operating parameters associated with displacement development. It also has been shown that TLC is an expedient concomittant to high performance displacement chromatography not only for the analysis of the column effluent but also in scouting for appropriate chromatographic systems and operating conditions (2,9).

Since then it has been shown that the use of TLC in the displacement mode, DTLC, can give rise to a workable separation method on its own in certain applications (2,9). In this report we present the results of an investigation concerning the effect of operating parameters on separations in DTLC.

EXPERIMENTAL

Materials:

TLC plates pre-coated with silica gel 60 F 254 were purchased from Merck (Darmstadt, FRG). Both 50 x 100 and 200 x 200 mm plates were used.

Monoethanolamine, diethanolamine, triethanolamine, chloroform, dichloromethane, dichloroethane, acetone, ethanol and ninhydrin were supplied by Reanal (Budapest, Hungary). Benzylamine, phenylethylamine, tyramine, dopamine and p-methoxyphenylethylamine, all hydrochlorides, were purchased from Fluka (Buchs, Switzerland). Deprenyl (N-methyl-N-propargyl-(2-phenyl-1-methyl) ethylammonium chloride, Jumex^{R}), propargylanara (N-propargyl-(2-phenyl-1-methyl)ethyl-ammonium chloride), amphetamine and methamphetamine were the kind gift of Chinoin (Budapest, Hungary).

The sudan black dye-mixture was purchased from Camag (Muttenz, Switzerland) under the name ."Test-Substance II" and used in carrier displacement TLC.

Apparatus:

Development chambers for 200 x 200 mm plates were from Desaga (Heidelberg, FRG). For development of 50 x 100 mm plates, 400 ml covered glass beakers of Symax (Prague, Czechoslovakia) were used. The solutions were spotted with µl syringes supplied by Scientific Glass Engineering (North Melbourne, Australia). A Minuvis UV lamp of Desaga was used at 254 nm for visualizing certain spots.

Methods:

After applying the sample solution at a distance of 30 mm from the lower edge of the TLC plate, the spots were dried by the use of cold air stream. The developer was allowed to rise until its front was 20 mm from the top of the plate. Thus the development distance on the 200 x 200 mm plates was 150 mm unless otherwise specified.

When the chamber was presaturated by the vapors of the developer solvent, a piece of filter paper was impregnated with the liquid and placed into the chamber an hour before start. In most cases solvent impregnated filter paper was not used and this type of development is referred to as no or without presaturation. All solutions were placed into the chamber at least two hours prior to the start of development.

The spots on the chromatogram were visualized either by the use of UV lamp or by spraying with a solution containing 0.2 w/v % ninhydrin in ethanol and subsequent heating at 110 ^{o}C for 10 minutes.

In carrier displacement TLC, the dye-mixture was spotted after the samples of substances to be separated had been dried by cold air stream. The sudan black components of the mixture did not show any mobility in chloroform, dichloroethane or dichloromethane but could be displaced by mono- or triethanolamines. The separated dye-components and phenylalkylamines displaced by the individual dye-components were directly observed visually without UV lamp or ninhydrin.

RESULTS AND DISCUSSIONS

In the first set of experiments we investigated the effect of presaturation by the solvent of the developer on the migration of the displacer front. The latter is conveniently measured by the R_D value that is obtained by dividing the distance of the displacer front from the bottom of the plate by the corresponding distance appropriate for the carrier front (9). In all experiments the R_D values measured without presaturation were higher than those with presaturation. For instance with chloroform and 5 % triethanolamine in chloroform as the carrier and displacer, respectively, the R_D value was 0.32 and 0.44 with and without presaturation, respectively.

Thus, in DTLC the migration of the displacer relative to that of carrier can be controlled not only by the nature and concentration of the displacer but also by the level of plate saturation with the vapors of developer. Since in DTLC high R_D values are generally desirable, the use of unsaturated chamber is recommended. Moreover heat activation of the plate may further augment the effect. Enhancement of displacer migration relative to carrier migration stems from the fact that the movement of the carrier front in unsaturated chamber is slower than in the presaturated one as illustrated in Fig. 1. The volatile carrier solvent is believed to evaporate partially from the plate when the chamber is not saturated with its vapor and consequently the solvent front migrates slower than when the plate is developed in the vapor-saturated chamber. On the other hand, the displacer is non-volatile and therefore its movement is only slightly affected by the level of saturation as it is seen in Fig. 1. Moreover, the R_D value is sufficiently low to be strongly influenced by the initial conditions of the vapor phase

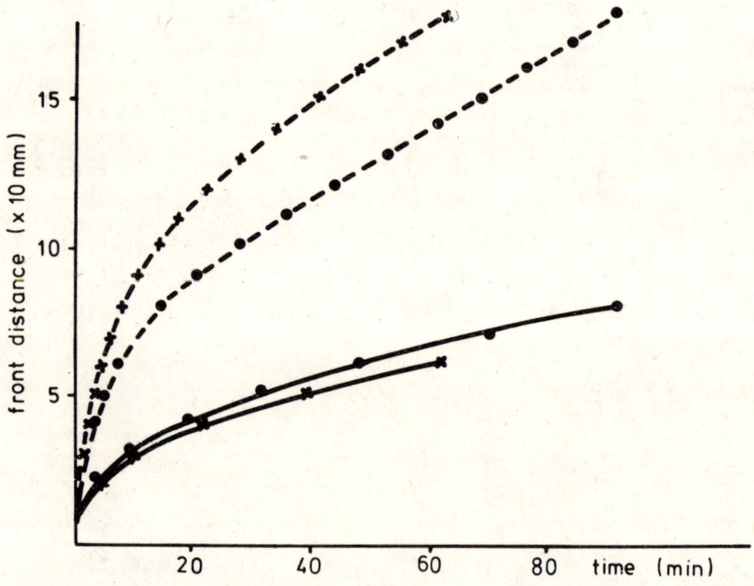

Figure 1. Migration of the displacer (solid line) and the carrier front (dashed line) on plates developed in chambers with (x—x—x) and without (●—●—●) presaturation with the chloroform used as carrier. The displacer was 5 % (v/v) triethanolamine in chloroform.

in the development chamber. Preactivation of the plate, however, may reduce the R_D value.

In another set of experiments the effect of the position of the sample spot relative to the displacer front was investigated in carrier DTLC. As illustrated in Fig. 2 the sample position with respect to the bottom of the plate was varied within wide limits. It is seen in Fig. 2 that complete separation of deprenyl and phenylethylamine was accomplished by using the components of the dye-mixture as carriers when the displacer front migrated a distance of 15 mm from the level of the initial spot. A closer distance resulted in incomplete displacement and of course the sample spot not reached by the displacer front did not exhibit any movement under such conditions as shown in Fig. 2. Each sample spot contained 25 μg phenylethylamine and 50 μg deprenyl. Phenylethylamine was displaced by both dye III and displacer front, deprenyl was displaced by both dye II and III as well as phenylethylamine and the displacer front, too, if the fully developed displacement train had been

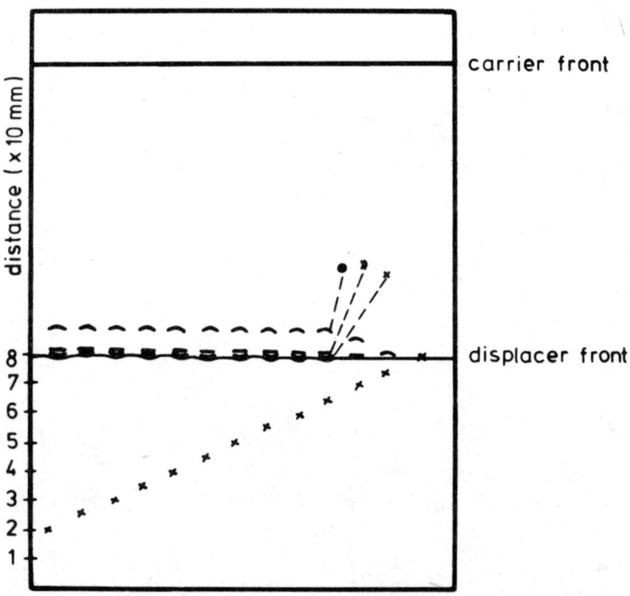

Figure 2. Effect of the position of the displacer front relative to the initial sample spots on the carrier displacement separation of deprenyl and phenylethylamine by using the components of sudan black (✗, ➤ and ● are dye I, dye II and dye III components, respectively). The chromatographic system is the same as that given in Fig. 1 for the chamber without presaturation.

built up. This situation prevailed in all cases except in the three spots on the right side where the length of the run was less than 15 mm.

These observations suggest that with the above mentioned limits the displacement train was fully developed under such conditions. Thus, for the separation of a simple mixture containing a few components, rather short plates suffice, as the displacer front need not migrate a long distance past the sample spots in order to obtain a fully developed displacement train that occupies a relatively short distance ahead of the front. In practical applications the sum of the displacer front and displacer train distances is quite short and in effect responsible for the reduction in the required plate length.

It is an eminent feature of displacement chromatography that it can be performed with sample spots much greater in diameter than those appropriate in elution mode. This is illustrated in Fig. 3 that shows no deterioration in separation due to a six-fold increase in the spot area and sample amount that was already much greater for the smallest sample than usual in elution TLC. It is noteworthy that the increase in the length of the sample spot had no untoward effect either and the separation was good even for the components initially present in a 30 mm long sample strip.

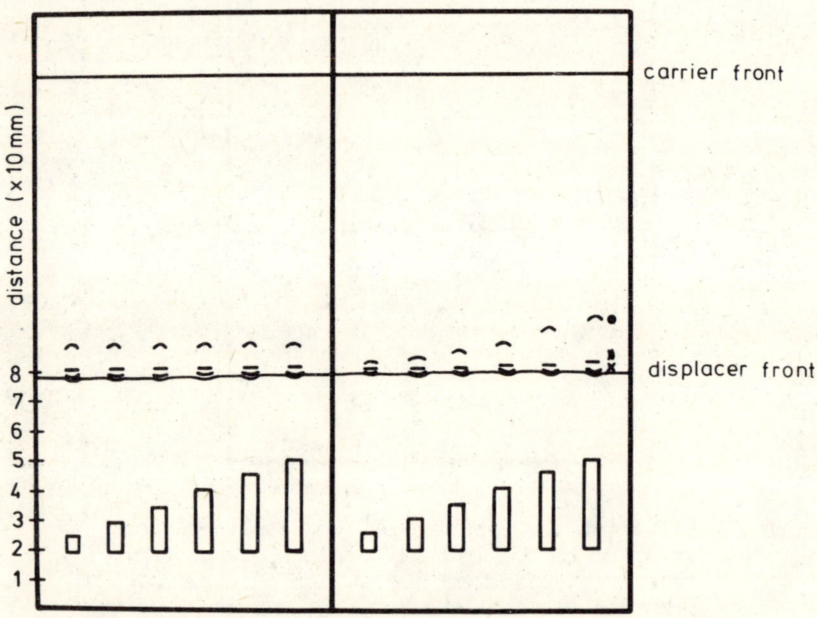

Figure 3. Effect of the sample load in carrier displacement chromatography. Conditions are the same as indicated in Fig. 2, excepting the sample places

On the left side of Fig. 3 each sample strip has the same total load (sample places are the columns as marked) that is 25 μg phenylethylamine and 50 μg deprenyl. It corresponds to 1 and 2; 1/2 and 1; 1/3 and 2/3; 1/4 and 2/4; 1/5 and 2/5; 1/6 and 2/6 $\mu g/mm^2$ phenylethylamine and deprenyl, respectively. On the right side of Fig. 3 each sample strip has the same specific load on square mm but increasing amounts in respect of the total load. 1/3 $\mu g/mm^2$ phenylethylamine and 2/3 $\mu g/mm^2$ deprenyl were spotted, namely the total amount of 8 and 16; 17 and 34; 25 and 50; 33 and 66; 42 and 84; 50 and 100 μg phenylethylamine and deprenyl, respectively, for the spots from left to right. Dye I, dye II and dye III components are marked with ", >> and *, respectively.

It is a particular feature of displacement chromatography, in contradistinction to elution, that the sample may be concentrated in the course of development depending on the adsorption isotherms of the displacer and/or of the sample components (6). The practical consequence of this manifests itself not only in a possible enhancement of sensitivity under certain conditions (2) but also in the relaxation of the usual requirement in TLC, that is, the extent of the sample spot is kept at minimum. This is strikingly illustrated in Fig. 3. Thus, dilute samples can be applied as several centimeter long strips in the direction of flow without adverse effect on the quality of separation. Such a compression effect may be of importance when the solubility of the components in the sample solvent is low or in trace analysis and preparative separations by DTLC. Of course, elongated sample spots call for a greater migration distance of the displacer front than that required under usual conditions.

We have investigated the effect of spotting the sample mixture and that of the carrier substances, i.e., the dye-mixture at different positions in the direction of flow on the plate. The carrier displacement separation was the same even when the two spots were spaced 20 mm apart and the order of spots had no observable effect either as shown in Fig. 4. In practice, therefore, no difficulties are expected when the solvents of the sample proper and of the carrier mixture are immiscible, e.g., when aqueous urine samples and solution of the dye-mixture in hexane have to be applied.

Selection of the carrier solvent and the carrier has a strong influence on the time and efficiency of separation by DTLC. In the experiments discussed above chloroform and 5 % triethanolamine in chloroform were used as the carrier and the displacer, respectively. However, other systems have

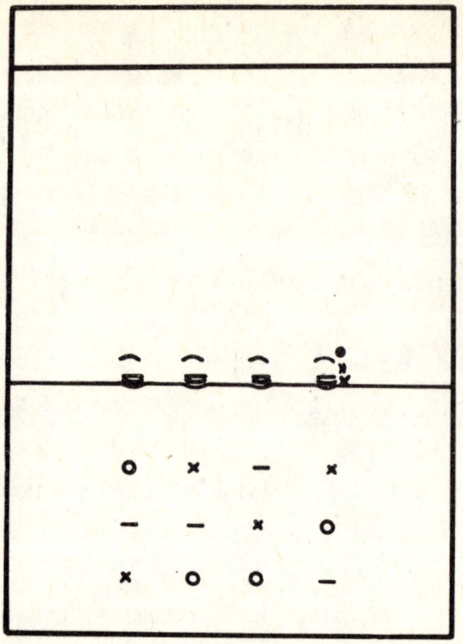

Figure 4. The ineffectiveness of location and order of the sample spots at the start of DTLC. Dye mixture (-), 50 µg deprenyl (×) and 50 µg phenylethylamine (o) were spotted at the places as indicated. After the displacement train had been fully developed, the spots were situated in their right order.

also been investigated and the results are shown in Table 1. It is seen that for the separation problem at hand various chlorinated aliphatic hydrocarbons and ethanolamines are suitable carriers and displacers, respectively. In general the data show that shorter development time results in shorter displacement train.

In our earlier reports, DTLC of phenylalkyl compounds was restricted to certain types (2, 13) because only non-hydroxy substituted phenylalkylamines were displaced by triethanolamine in chloroform or in several other systems containing chlorinated hydrocarbons as the carrier and alkylamines as the displacers. Although a variety of other alkylamines were also tested as potential displacers they gave similar results. However, most recent investigation showed that by changing the composition of the carrier solvent it is possible to displace hydroxy-substituted phenylalkylamines, too. Thus we found that p-tyramine can be displaced by a system consisting of acetone-chloroform (1:1) and 5 % triethanolamine as the carrier and displacer, respectively.

Table I.

Carrier	Displacer	%	Running time	R_f of the Displacement 2nd front	range (mm)
chloroform	ethanolamine	5	41	28/80	10
		10	43	50/80	2
		20	51	67/80	2
	triethanolamine	5	26	30/80	10
		10	x	67/69	**
		20	x	62/63	2
dichloromethane	ethanolamine	5	23	23/80	10
		10	27	30/80	2
		20	x	68/70	2
	triethanolamine	5	20	30/80	10
		10	28	48/80	2
		20	x	54/53	2
dichloroethane	ethanolamine	5	27	6/80	?
		10	x	6/15	?
		20	36	42/80	2
	triethanolamine	5	29	16/80	9
		10	29	24/80	5
		20	x	25/40	4
chloroform-acetone (1:1)	ethanolamine	5	26	45/80	4
		10	49	76/80	2
		20	66	77/79	**
	triethanolamine	5	19	23/80	10
		10	28	35/80	3
		20	38	54/80	2

x : time of development is more than 90 minutes for 80 mm

** : displacement range is less than 1 mm

? : displacement range is unknown (as the 2nd front does not reach the origin of spots).

Table II.

	R_f x 100			
	I	II	III	IV
Benzylamine	34	96	43	95
Phenyethylamine	34	96	43	95
Tyramine	Ø	Ø	42	93
Dopamine	Ø	Ø	Ø	19
Amphetamine	35	97	44	95
Methamphetamine	35	97	44	95
Deprenyl	36	97	78	98
Propargylanara	37	98	82	98
p-methoxyphenyl-ethylamine	Ø	Ø	Ø	28

Solvent systems:

I: 5 % triethanolamine in chloroform

II: 10 % triethanolamine in chloroform

III: 5 % triethanolamine in chloroform-acetone (1:1)

IV: 5 % ethanolamine in chloroform-acetone (1:1).

The results are shown in Table 2, and it is seen that the spot of p-tyramine is displaced between the displacer front and phenylethylamine, and amphetamine and methamphetamine are also displaced. On the other hand, deprenyl and propargylanara elute when acetone-chloroform (1:1) is used as the carrier that appears to be a strong eluent due to the presence of the highly polar acetone. With acetone-chloroform mixture as the carrier, triethanolamine is also capable of displacing tyramine, which too strongly binds to silica gel in the presence of chloroform only to be eluted or displaced by alkylamines, so that its spot appears directly ahead of the displacer front.

The results presented in Table 2 show that the ethanolamines investigated have similar properties as displacers as long as a chlorinated solvent is used as the carrier. Only the addition of acetone to the carrier gives rise to significant change in the chromatographic behaviour of phenylalkylamines. Whereas phenylethylamine, benzylamine, amphetamine, methamphetamine are still displaced under such conditions, *vide* Table 2, deprenyl and propargylanara are eluted and dihydroxy derivatives do not move on the thin-layer plate at all.

The length of migration distance plays an important role in determining the time and efficiency of separation in DTLC. In analogy with recent findings in column chromatography (7), the process can be broken down in the following steps:

(a) displacer front migrates prior reaching the spots of the sample mixture;

(b) displacer front is touching the spots of sample and concentrates the components even if they are present in more or less elongated strips;

(c) the components start to move and form a displacement train of adjacent spots upon the effect of moving displacer front;

(d) after reaching isotachic conditions, all bands of the fully developed displacement train move with the velocity of the displacer front.

The actual separation process takes place in steps (b) and (c) only. Therefore, the time and space used for steps (a) and (d) should be minimized. Upon reaching isotachic conditions, the separation in the fully developed displacement train may deteriorate due to diffusion and flow effects. Consequently, the development in DTLC should be stopped after step (c). Thus, the maximum migration distance for the sample on the plate is the sum of the distance needed by the displacement train. As seen in Fig. 2 both distances can be rather short. Consequently, the use of 100 mm high thin-layer plates appear to be quite adequate in our case. The situation contrasts with that in elution TLC where the disengagment of spots is generally proportional to the square root of the migration distance (10-12).

REFERENCES

(1) STAHL,E. (Dünnschicht-Chromatographie. Springer, Berlin, 1967.

(2) KALÁSZ,H., NAGY, J., KNOLL, J. in(FRIGERIO,A. (Editor): Chromatography and Mass Spectrometry in Biomedical Sciences,2. pp. 203-214. Elsevier, Amsterdam, 1983.

(3) TISELIUS,A.: Ark. Kemi, Mineral. Geol. 16A 1 (1943)

(4) PORATH,J.: Acta Chem.Scand. 6 1237 (1952)

(5) CLAESON,S.: Ann. N.Y. Acad.Sci. 49 183 (1948)

(6) HORVÁTH,CS., NAHUM,A., and FRENZ,J.: J.Chromatogr. 218 365 (1981)

(7) HORVÁTH,CS., FRENZ,J., and EL RASSI,Z.: J.Chromatogr. 255 273 (1982)

(8) KALÁSZ,H., and HORVÁTH,CS.: J. Chromatogr. 215 295 (1981)

(9) KALÁSZ,H., and HORVÁTH,CS.: J.Chromatogr. 239 423 (1982)

(10) FENIMORE,D.C., in BERTSCH,W., HARA,S., KAISER,R.E. and ZLATKIS,A. (Editors): Instrumental HPTLC. Hüthig, Heidelberg, 1980.

(11) GIDDINGS,J.C. in HEFTMANN,E. (Editor) Chromatography, 3rd.Edition. Reinhold, New York, 1975.

(12) VERNIN,G.: La chromatographie en couche mince. Dunod, Paris, 1970.

(13) KALÁSZ,H.: J. High Resol. Chromatogr. Chrom. Communic. 6 49 (1983)

CHARACTERIZATION OF LIQUID CHROMATOGRAPHIC STATIONARY PHASES

Proc. of the Symp. on Advances in Liquid Chromatogr., Szeged, Hungary, 1982

"CHROMSIL". A NEW FAMILY OF CHROMATOGRAPHIC PACKINGS

R. OHMACHT,* Z. MATUS**

*Institute of Chemistry, University Medical School, Pécs, Hungary
**"Labor" Instrument Works, Budapest, Hungary

SUMMARY

The Chromsil, Chromsil-ODS and Chromsil-NH_2 packings were systematically examined. It was found that pore structures of these materials are optimum. The efficiency and the packing homogenity of packed columns were characterized by H-u curves. Several applications of these packings are presented.

INTRODUCTION

High performance liquid chromatography (HPLC) is one of the most important analytical methods in chemistry and biochemistry. The types of special packing materials developed for HPLC have been multiplied during the last two decades. The most common material is silica. It can be used in two different ways as packing: as an adsorbing surface of silanol groups or through other polar, nonpolar or ion-exchanger organic groups attached to the surface. The particles should be of uniform size to ensure uniform flow across a column cross-section and to obtain maximum efficiency within the limits of practical pressure drop through the column and with a uniform porosity and optimum structure for the particulate in use.

The enormous progress of HPLC in the last decade can be partly attributed to the introduction of chemically bonded phases (1,2). Attachment of various polar and nonpolar groups to the surface by direct Si-C bonds generally gives a stable coating toward hydrolysis. The steadily increasing number of papers demonstrates the growing importance of these separation systems.

The "Chromsil" packings for HPLC and high performance thin-layer chromatography are based on a new, irregular shaped silica. They are available in 6 µm and 10 µm average particle size. The efficiency of this material and its derivatives, the chemically bonded "Chromsil-ODS" and "Chromsil-NH_2" was investigated by measuring the porosity and H-u curves.

These packing materials have been developed by the University Medical School, Pécs in collaboration with the "Labor" Instrument Works, Budapest.

EXPERIMENTAL

The following liquid chromatograph was used: Liquochrom Model 2010 (Labor MIM, Budapest, Hungary) consisting of a pulseless solvent delivery pump, an injector valve with 20 µl loop, a variable wavelength UV detector and a strip-chart recorder, Type OH-814/1 (Radelkis, Budapest, Hungary).

The packing materials were filled into 250 x 4.6 mm stainless steel columns Type OE-323 (Labor MIM, Budapest, Hungary). The packing procedure based on a modified "balanced-viscosity" method (3) has been specially developed by Labor MIM. Before use, the packed columns were equilibrated with the eluent for an hour.

The components of all mobile phases were freshly distilled. The test-mixtures were prepared by mixing analytical grade standard materials.

RESULTS AND DISCUSSION

1) Silica packing material

The pore size distribution, pore volume, specific surface area have been measured by size-exclusion chromatography (4). The average pore diameter was found at $\emptyset$ = 83 Å (Fig.1, solid line), the pore volume was $V_{p,sp}$ = 1.07 cm^3/g, the specific surface area was S = 520 m^2/g.

This value of surface area measured chromatographically is usually 30 % higher than measured by BET method.

The total porosity, the pore porosity and the interstitial porosity were $\varepsilon_T = 0.82$. $\varepsilon_p = 0.40$ and $\varepsilon_o = 0.42$, respectively.

The efficiency of the columns can be characterized by the van Deemter equation (5):

$$H = A + \frac{B}{u} + C\,u$$

where H is the high equivalent to a theoretical plate, u is the linear flow velocity, A, B and C are characteristics to the column, packing material, solvent and solute.

We can use the complete form of this equation for columns packed with materials having about 5 µm diameter only. We have characterised the 10 µm materials with a simplified van Deemter equation:

$$H = A^* + C^*\,u$$

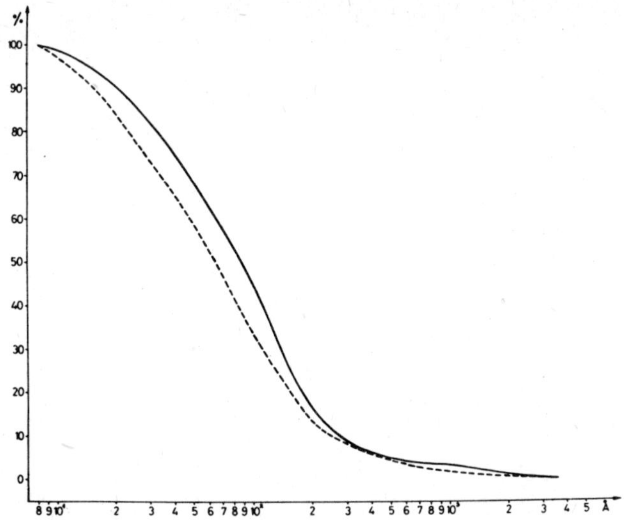

Fig.1 Pore size distribution of silica (solid line) and C_{18} bonded phase (dashed line) packings

Heptane eluent, and tetrachloroethylene (k' = 0), anthracene(k'~ 1.0), 3,4-benzopyrene (k'~2.0) solutes were used. To indicate the packing homogeneity in the columns, the H versus u curves were measured not only in the direction of packing but with inverted flow direction, too. The results (Figs. 2,3,4) show that the efficiency of columns are nearly identical in both directions of flow.

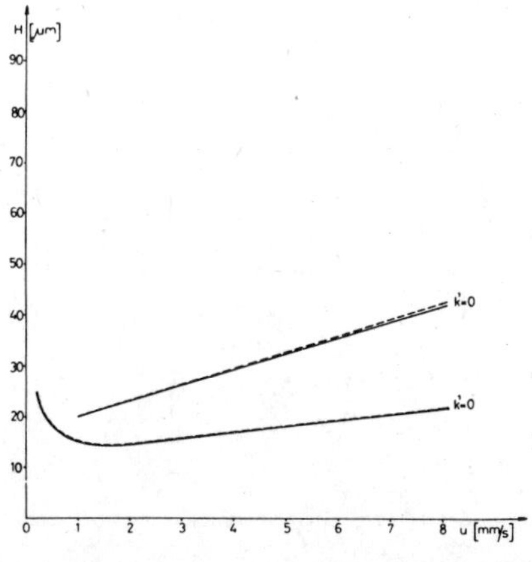

Fig.2. H-u curves on silica at k'=0 (Perchloroethylene sample)
Eluent: n-heptan
Detector: 254 nm; 0,1 A.
Upper lines: column with 10 μm packing material
Lower lines: column with 6 μm packing material
Solid lines: measured by flow direction corresponding to packing direction
Dashed lines: measured by inverted flow direction

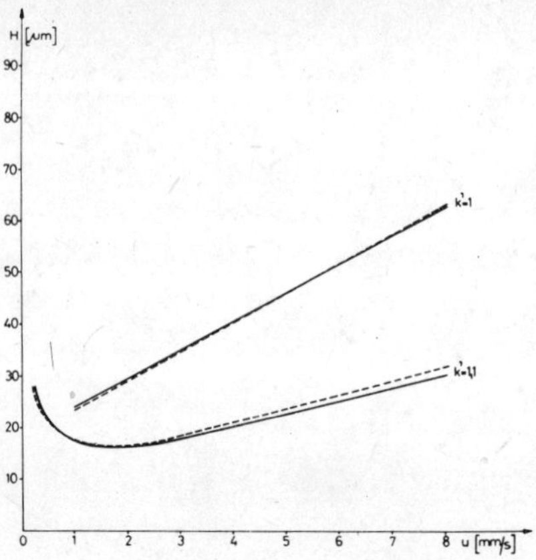

Fig.3. H-u curves on silica at k'~ 1 (Anthracene sample)
For details see Fig.2.

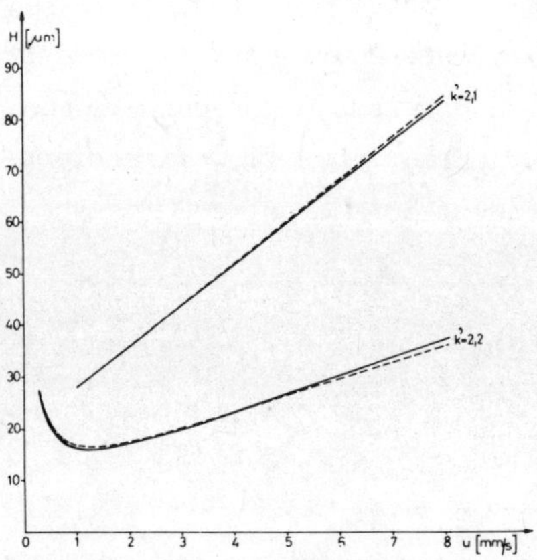

Fig.4. H-u curves on silica at k'~ 2(3,4-Benzopyrene sample)
For details see in Fig.2.

The reduced plate height at the optimum "u" value is about 2.5.

Fig.5 shows a size exclusion chromatogram of a mixture of polystyrenes with different molecular weights, on a 6 μm Chromsil column. The eluent was dichloromethane.

For the separation of some important hormones a 10 um Chromsil column and chloroform-methanol eluent were used (Fig.6).

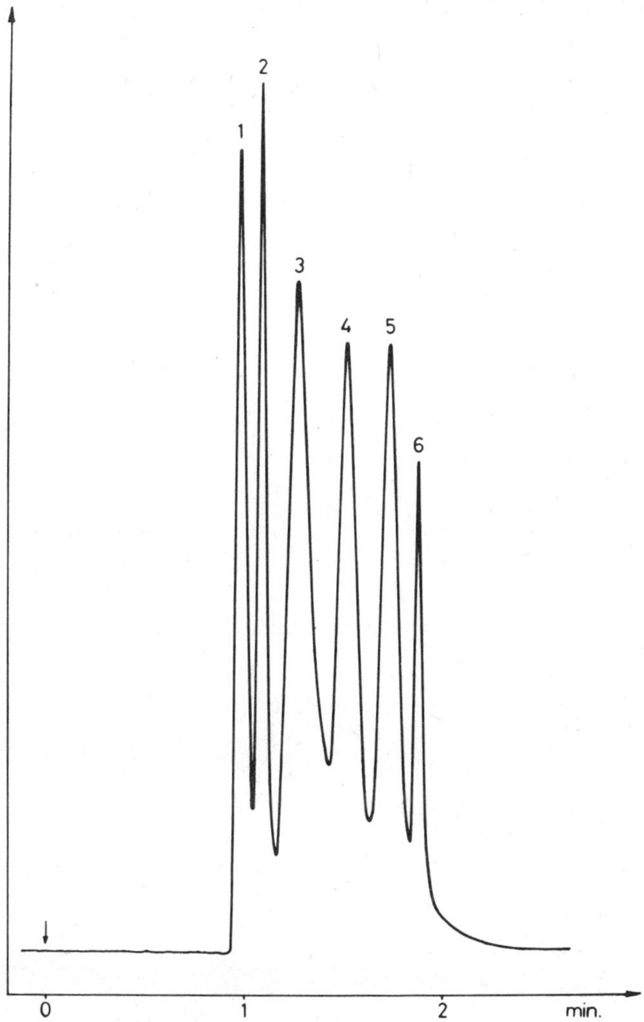

Fig.5. Separation of polystyrene samples by size exclusion.
Packing material: 6 µm Chromsil
Mobile phase: dichloro-methane
Linear velocity: 2,23 mm/sec
Detector: 260 nm; 0,1 A.
Peaks: 1. Polystyrene M.W. 3 700 000
2. Polystyrene M.W. 20 000
3. Polystyrene M.W. 10 000
4. Polystyrene M.W. 3 000
5. Polystyrene M.W. 600
6. Benzene 78

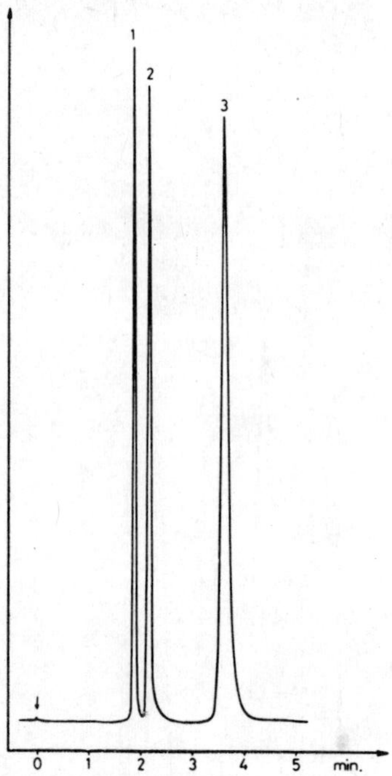

Fig.6. Separation of some steroid-hormones
Packing material: 10 μm Chromsil
Mobile phase: $CHCl_3$: MeOH 97:3 (v:v)
Linear velocity: 2,40 mm/sec
Detector: 280 nm; 0,1 A.
Peaks: 1. Estrole
2. Estradiol
3. Estriol

2) Chromsil-ODS packing material

This reverse phase packing material is based upon the interaction of Chromsil silica with trichlorooctadecylsilane. The surface is well capped and has a very low concentration of the residual silanol groups. The carbon content is 17.6 %. The porosity data were measured in a similar way to silica. The results are: mean pore diameter, pore volume and specific surface area are 63 Å (Fig.1, dashed line), 0.58 cm^3/g and 370 m^2/g, respectively, ε_T = 0.70; ε_p = 0.29; ε_o = 0.41. All data, excepting ε_o, decrease, because the volume of octadecyl chains can not be neglected.

The efficiency of columns was tested in two different ways. First, we measured the H vs. u curves in methanol eluent benzene (k' = 0.2), pyrene (k' = 1) and 3,4-benzpyrene (k' = 2.1) in both flow directions (Figs.7,8 and 9). Then, we changed the eluent to 89 % water and 11 % methanol. In this case, nitromethane (k'$\sim$ 0.6), hydroquinone (k'$\sim$1.0) and resorcinol(k'$\sim$2.3) solutions were injected (Fig.10).

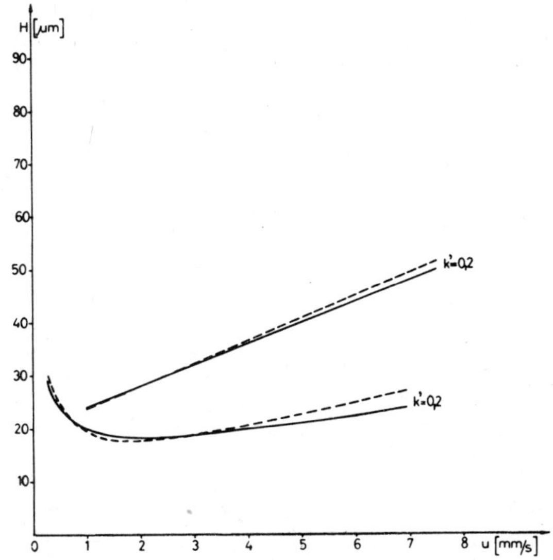

Fig.7. H-u curves on Chromsil-ODS at k' = 0,2 (Benzene sample)
Eluent: methanol
Detector: 254 nm; 0,1 A.
Upper lines: column with 10 µm packing material
Lower lines: column with 6 µm packing material
Solid lines: measured by flow direction corresponding to packing direction
Dashed lines: measured by inverted flow direction

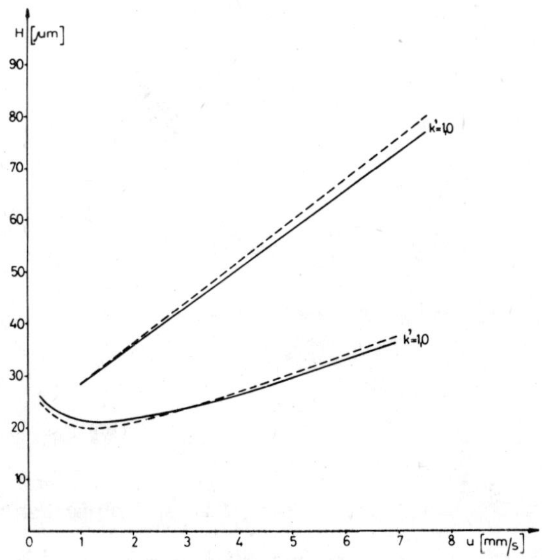

Fig.8. H-u curves on Chromsil-ODS at k' = 1 (Pyrene sample)
For details see Fig.7.

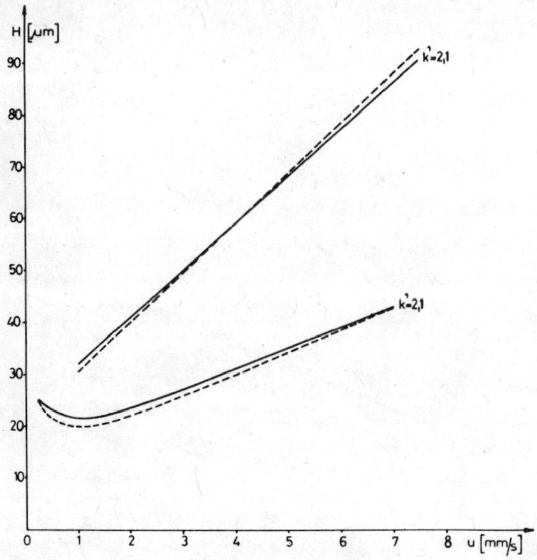

Fig.9. H-u curves on Chromsil-ODS at k' = 2,1 (3,4-Benzopyrene sample)
For details see Fig.7.

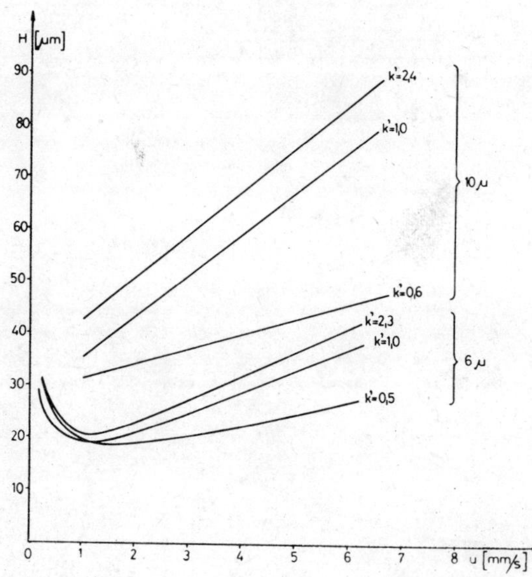

Fig.10. H-u curves on Chromsil-ODS with water-methanol eluent.
Detector: 254 nm; 0,1 A.
Mobile phase: 11 % methanol in water
Samples: Nitromethane (k'~ 0,5), Hidroquinone(k'=1,0),
Resorcinol (k'~2,3)

The most important difference between the two methods is that the stationary phase (the ODS-silica) has been wetted by methanol and not by water--methanol (89:11) eluent. The results in both eluent are very similar to each other.

In Figs. 11 and 12 two applications of "Chromsil-ODS" are presented. Fig. 11 demonstrates a rapid separation of some halogenated benzenes on a column packed with 6 um particles. In Fig.12 there is a separation of an alkaloid-cocktail.

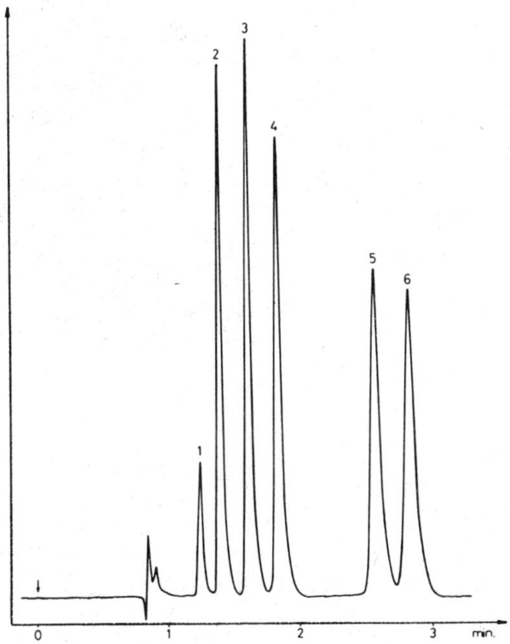

Fig.11. Separation of some halogeneted benzenes
Packing material: 6 μm Chromsil-ODS
Mobile phase: MeOH:water 88:12 (v:v)
Linear velocity: 4,97 mm/sec
Detector: 270 nm; 0,05 A.
Peaks: 1. Beneze
2. chloro-benzene
3. 1,4-dichloro-benzene
4. 1,4-dibromo-benzene
5. 1,2,3-tri-chloro-benzene
6. 1,2,3,4-tetrachloro-benzene

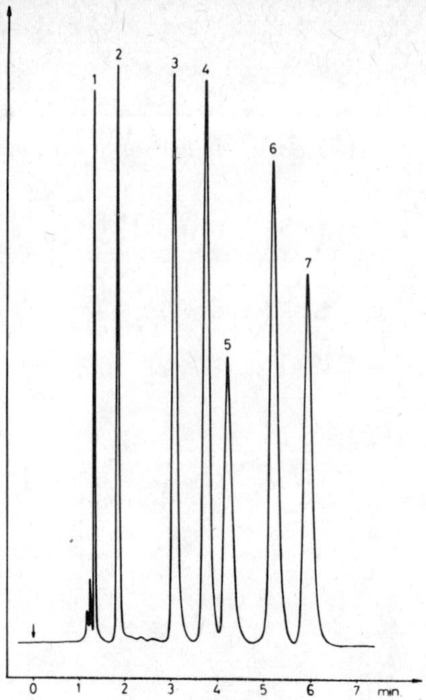

Fig.12. Separation of alkaloids
Packing material: 10 um Chromsil-ODS
Mobile phase: 55 % water, 44,7 % acetonitrile, 0,3 % diethylamine
Linear velocity: 3,50 mm/sec
Detector: 258 nm; 0,1 A.
Peaks: 1. Morphine
2. Caffeine
3. Nicotine
4. Papaverine
5. Strychnine
6. Thebaine
7. Cocaine

3) Chromsil-NH_2

This packing has chemically bonded amino-propyl groups on its surface. The residual silanol groups were eliminated by capping procedure, so acidic properties can not be observed.

For measuring H-u curves benzene (k' = 0), m-dinitrobenzene(k'=0.6) and o-nitroaniline (k' = 1.9) solutes were used. The eluent contained 0.5% i-propanol and 10 % dichloromethane in n-heptane (Fig.13).

Aminophase packings are very frequently applied for carbohydrate analysis. We separated a test-mixture of five commonly used mono- and disaccharides (Fig.14).In this case, refractive index detector was applied.

We have found that aminophase packings are suitable for the separation of carotenoids. For example, a "functional group separation" of carotenoids is presented in Fig. 15. Peak 1 (β -carotene) is a hydrocarbon, peak 2. (β-cryptoxanthin) a monohydroxy-peak 3 (zeaxantin) a dihydroxy-, peak 4 (antheraxantin) a monoepoxy-dihydroxy-, peak 5 (capsanthin) a monooxy-dihydroxy-, and peak 6 (capsorubin) a dioxo-dihydroxy-carotenoid.

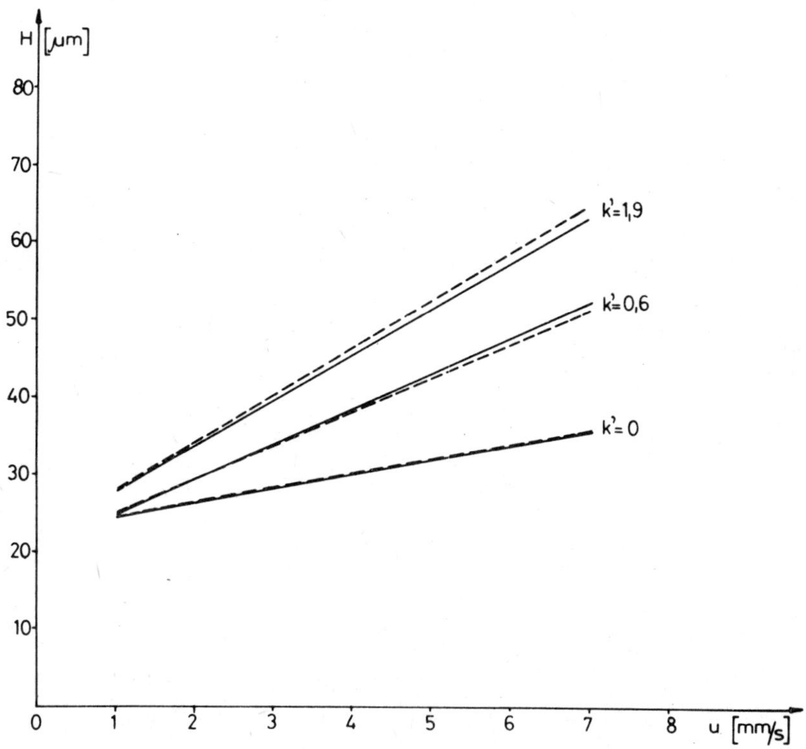

Fig.13. H-u curves on Chromsil-NH_2 (10 μm) phase
Mobile phase: 0,5 % isopropanol, 10 % CH_2Cl_2 in n-heptane
Detector: 254 nm; 0,1 A.
Samples: Benzene (k'=0), 3,5-dinitro-toluene (k'=0,6), 2-nitro-aniline (k'=1,9)
Solid lines: measured by flow direction corresponding to packing direction
Dashed lines: measured by inverted flow direction

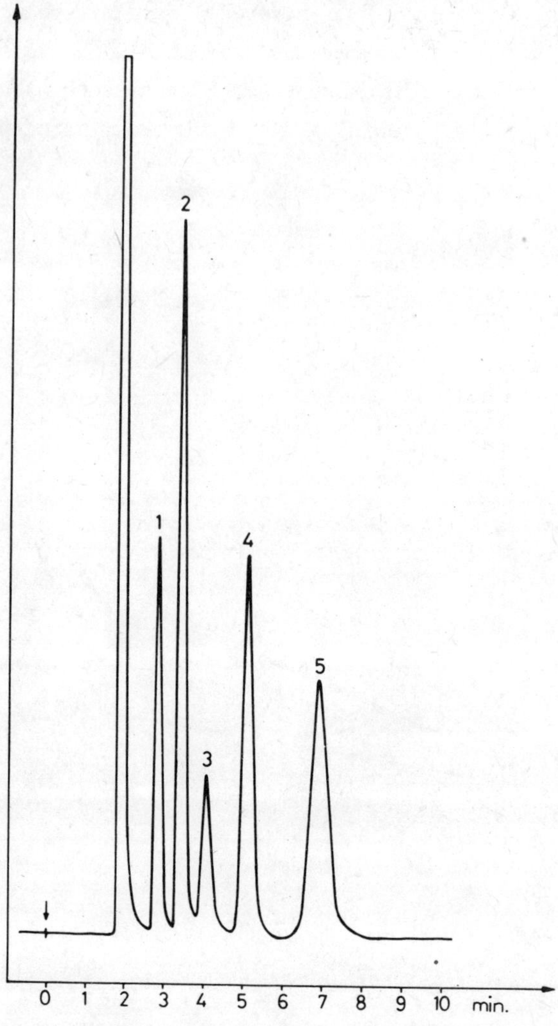

Fig.14. Separation of same mono- and disaccharides
Packing material: 10 µm Chromsil-NH_2
Mobile phase: 12 % water in acetonitrile
Linear velocity: 2,25 mm/sec
Detector: Refractive Index LCD
Peaks: 1. Xylose
2. Fructose
3. Galactose
4. Sucrose
5. Lactose

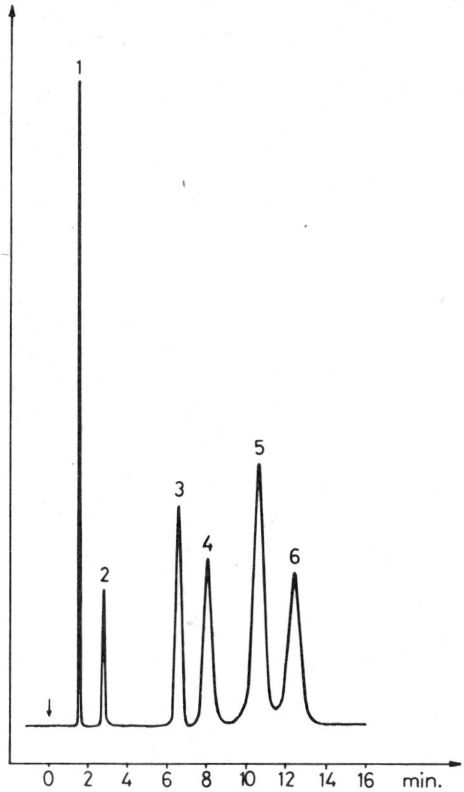

Fig.15. Separation of carotenoids
Packing material: 10 µm Chromsil-NH_2
Mobile phase: 0,5 % methanol in benzene
Linear velocity: 2,70 mm/sec
Detector: 470 nm; 0,08 A.
Peaks: 1. β-Carotene; 2. β-Cryptoxanthin; 3. Zeaxanthin; 4. Antheraxanthin; 5. Capsanthin; 6. Capsorubin

REFERENCES

(1) ENGELHARDT,H. High Performance Liquid Chromatography. Springer.Berlin, Heidelberg, New York, 1979
(2) WALTON,H.F. Anal.Chem. 52 15R (1980)
(3) OHMACHT,R. and HALÁSZ,I. Chromatographia 14 155 (1981)
(4) HALÁSZ,I., MARTIN,K. Angew.Chem.Int.Ed.Eng. 17 901 (1978)
(5) van DEEMTER,J.J. et al.: Chem.Eng.Sci. 5 271 (1936)

Proc. of the Symp. on Advances in Liquid Chromatogr., Szeged, Hungary, 1982

THE ROLE OF THE SPECIFIC SURFACE AREA OF AN ADSORBENT IN THE OPTIMIZATION OF MIXTURE SEPARATION CONDITIONS IN THIN-LAYER CHROMATOGRAPHY

J.K. ROZYLO, I. MALINOWSKA

Institute of Chemistry,
Maria Curie-Sklodowska University,
Lublin, Poland

Adsorbents of different microporous structure (1-5) have been used for a long time in thin-layer chromatography for the separation of mixtures of various substances. The microporous structure of adsorbents is an important feature affecting optimization of the chromatographic process. This was studied in thin-layer chromatography by Geiss (6) and Snyder (7). Practical possibilities of controlling the process of thin-layer chromatography were also indicated by Rozylo (8,9). There were also attempts to explain the physical-chemical significance of the value of the adsorbent specific surface area in thin-layer chromatography (10). The recent considerable interest in the role of the specific surface area of the adsorbent in thin-layer chromatography is due to the growing importance of this method as a pilot technique for determining the optimum conditions of mixture separation on both analytical and preparative scales. The latter is widely applied for obtaining and purifying of substances of natural origin in laboratories, as well as in pharmaceutical and food industries, and also in enviromental protection (11).

Linear relations between R_M values of substances for pure components of a mixed e.g. binary mobile phase $R_{M\ 1.2}$ and the specific surface area of the adsorbent, as has already been observed - are not favourable enough to be applied in a routine process of optimization of separation conditions. Therefore, the present paper analyses the relations between the R_M values of substances and the specific surface area of the adsorbent. The values were investigated for identical concentrations of the mobile phase on adsorbents of different values of the specific surface area.

MATERIAL AND METHODS

Measurements were taken of the R_F (R_M) values of model substances obtained in the process of adsorption thin-layer chromatography on silica gels of different microporous structure, produced by Merck (12).

pore diameter (Å)	specific surface area (m^2/g)
60	500
100	400
200	150
500	50

The chromatographed substances were certain polycyclic aromatic hydrocarbons showing neither electron-donor nor electron-acceptor properties (naphthalene, β-methylnaphthalene, pyrene, chrysene, fluoranthene, diphenyl). This selection of model chromatographed substances aimed at eliminating possible additional intermolecular actions with the adsorbent surface and the components of the mobile phase which was composed of two mixed binary solvents: benzene-methanol and benzene-ethanol. The procedure of conducting the chromatographic process and substance detection were described in several earlier publications (1-4,8, 13). The results are presented in tables with the R_M values of substances from the experiment and values calculated theoretically on the basis of the obtained correlations; graphs of the R_M values were also drawn, with respect to the value of the specific surface area of the adsorbent and the concentration of the more active component of the mixed mobile phase (1-3), as well as graphs of relations of the R_M values of substances in the function of concentration of the mobile phase (4-6).

DISCUSSION

It is apparent from the results presented in tables I-II and in Figures 1-5 that on silica gels of different specific surface areas the R_M values of the chromatographed substance

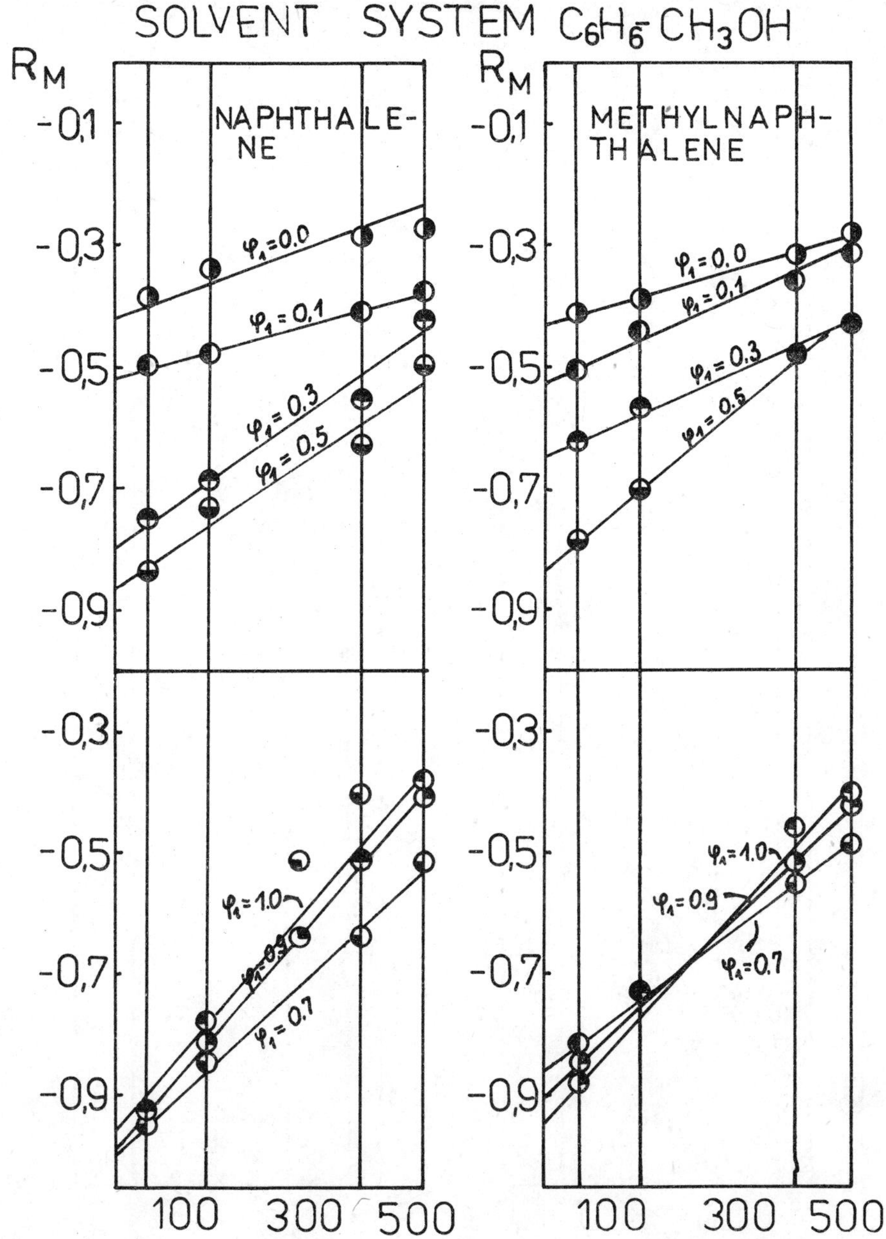

Figs1, 2, 3.

Linear relationship between R_M values and specific surface area of adsorbents (mg^2/g). Solvent systems. benzene-methanol. Experimental data, straight line-theoretical data, calculated from equation 1

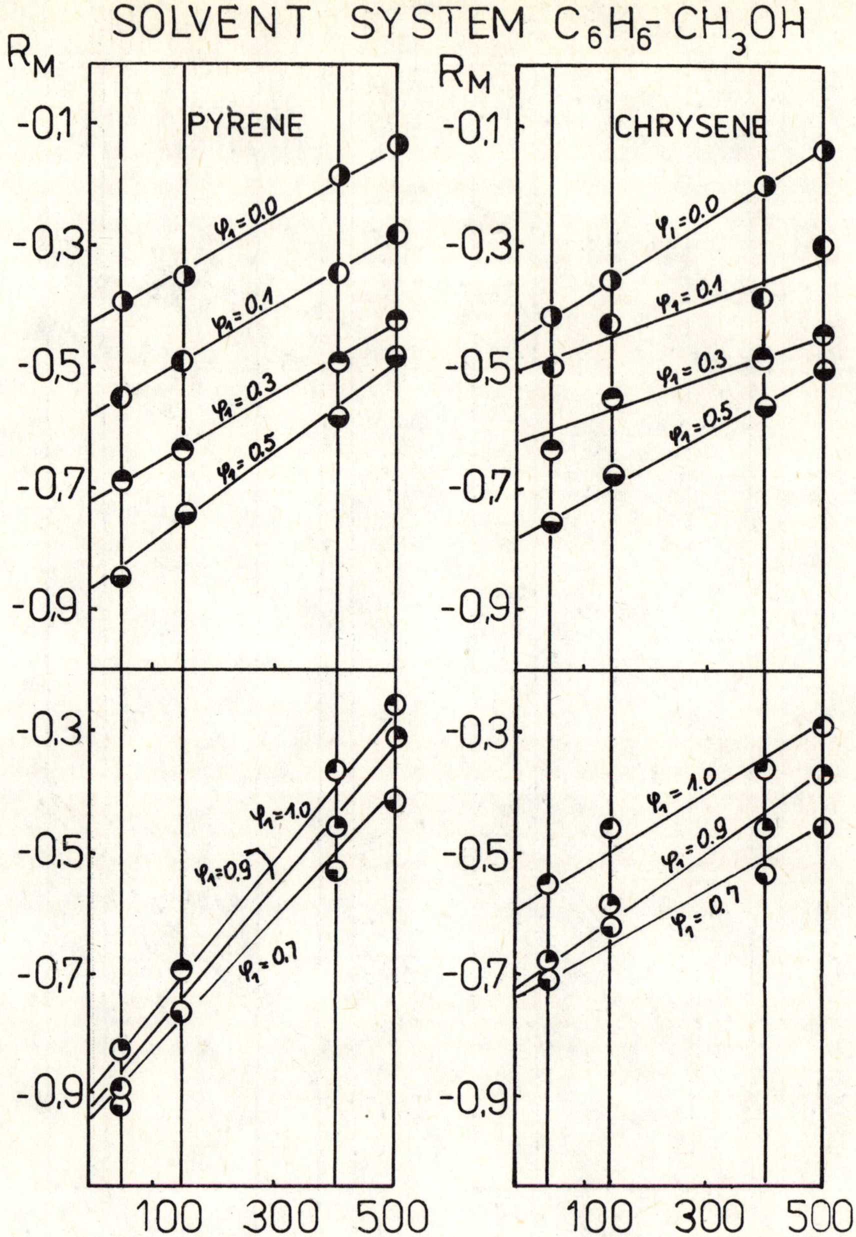
SOLVENT SYSTEM C_6H_6-CH_3OH
R_M
PYRENE
CHRYSENE
-0,1
-0,3
-0,5
-0,7
-0,9
$\varphi_1 = 0.0$
$\varphi_1 = 0.1$
$\varphi_1 = 0.3$
$\varphi_1 = 0.5$
$\varphi_1 = 1.0$
$\varphi_1 = 0.9$
$\varphi_1 = 0.7$
100
300
500

Fig. 2

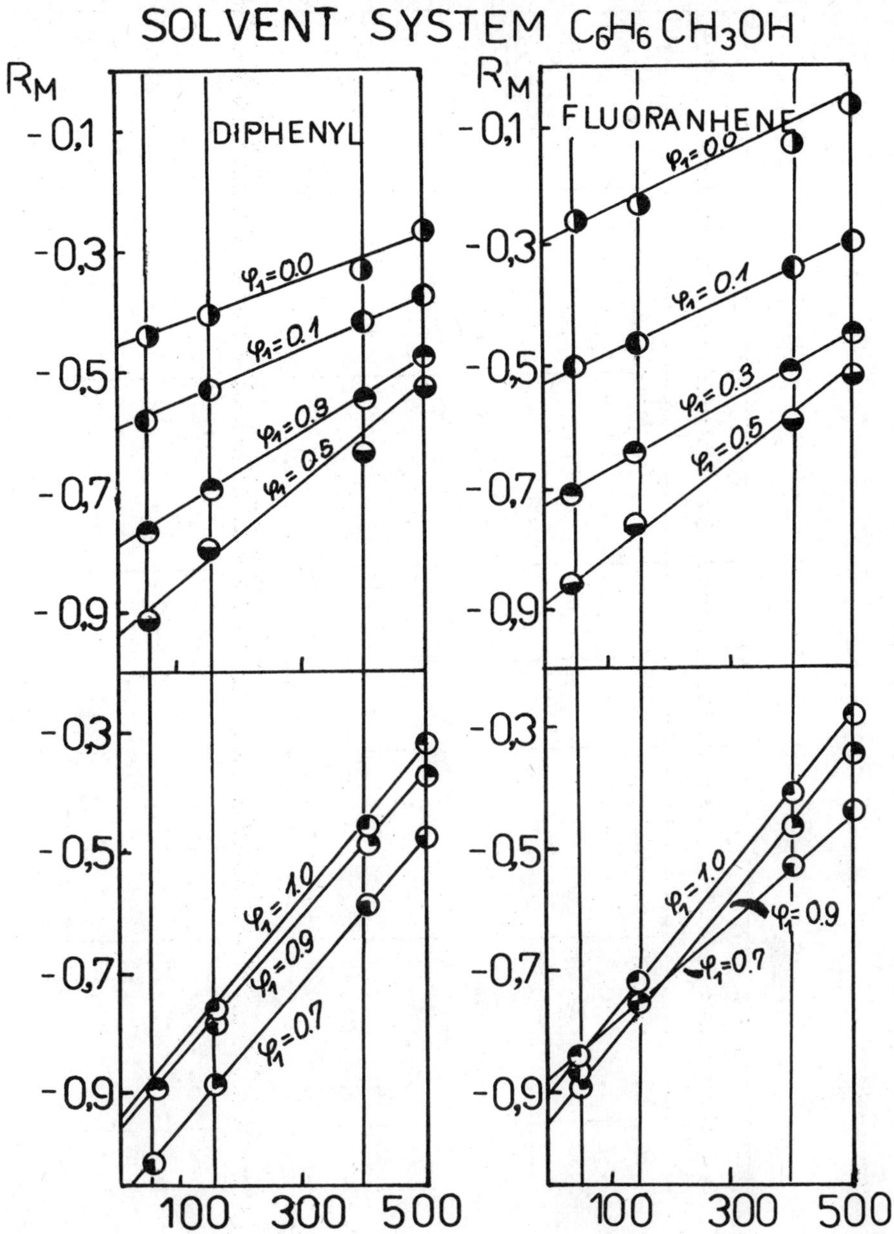
SOLVENT SYSTEM C6H6 CH3OH
RM
DIPHENYL
FLUORANHENE
φ1=0.0
φ1=0.1
φ1=0.3
φ1=0.5
φ1=1.0
φ1=0.9
φ1=0.7
-0,1
-0,3
-0,5
-0,7
-0,9
100
300
500

Fig. 3

TABLE I

Experimental and theoretical /calculated from equation 1/ R_M values for different chromatographed substances. Mobile phase benzene-methanol.

φ_{CH_3OH}	0,0		0,1		0,3		0,5		0,7		0,9		1,0	
	exp	calc	exp	calc	exp	calc	exp	calc	exp	calc	exp	calc	exp	calc
Naphthalene														
gel 60	-0,22	-0,24	-0,37	-0,37	-0,44	-0,42	-0,53	-0,52	-0,52	-0,52	-0,44	-0,41	-0,38	-0,35
gel 100	-0,26	-0,27	-0,40	-0,41	-0,51	-0,51	-0,60	-0,62	-0,62	-0,62	-0,55	-0,52	-0,51	-0,51
gel 200	-0,36	-0,34	-0,47	-0,46	-0,69	-0,69	-0,76	-0,75	-0,83	-0,75	-0,85	-0,84	-0,81	-0,80
gel 500	-0,39	-0,39	-0,50	-0,49	-0,76	-0,75	-0,82	-0,85	-0,92	-0,85	-0,96	-0,94	-0,95	-0,92
Methylnaphthalene														
gel 60	-0,28	-0,29	-0,30	-0,30	-0,42	-0,43	-0,43	-0,47	-0,48	-0,46	-0,41	-0,39	-0,38	-0,38
gel 100	-0,32	-0,32	-0,35	-0,35	-0,49	-0,48	-0,49	-0,54	-0,55	-0,52	-0,52	-0,47	-0,45	-0,45
gel 200	-0,38	-0,39	-0,44	-0,44	-0,56	-0,58	-0,71	-0,72	-0,72	-0,74	-0,74	-0,76	-0,74	-0,76
gel 500	-0,41	-0,41	-0,50	-0,50	-0,64	-0,62	-0,79	-0,78	-0,82	-0,86	-0,84	-0,88	-0,88	-0,88
Chrysene														
gel 60	-0,25	-0,25	-0,32	-0,30	-0,45	-0,44	-0,51	-0,51	-0,47	-0,47	-0,37	-0,38	-0,30	-0,30
gel 100	-0,28	-0,28	-0,36	-0,38	-0,49	-0,50	-0,58	-0,57	-0,52	-0,53	-0,44	-0,45	-0,35	-0,37
gel 200	-0,38	-0,36	-0,45	-0,43	-0,57	-0,56	-0,70	-0,67	-0,65	-0,63	-0,61	-0,60	-0,50	-0,48
gel 500	-0,42	-0,43	-0,49	-0,50	-0,60	-0,62	-0,76	-0,76	-0,71	-0,71	-0,69	-0,68	-0,55	-0,55

Table I / continous /

φ_{CH_3}	0,0		0,1		0,3		0,5		0,7		0,9		1,0	
	exp	calc	exp	calc	exp	calc	exp	calc	exp	calc	exp	calc	exp	calc
Pyrene														
gel 60	-0,25	-0,25	-0,50	-0,28	-0,50	-0,43	-0,48	-0,50	-0,41	-0,45	-0,33	-0,32	-0,24	-0,24
gel 100	-0,29	-0,28	-0,35	-0,35	-0,50	-0,49	-0,58	-0,58	-0,52	-0,55	-0,41	-0,43	-0,35	-0,37
gel 200	-0,36	-0,36	-0,45	-0,49	-0,64	-0,63	-0,74	-0,75	-0,77	-0,79	-0,72	-0,74	-0,69	-0,70
gel 500	-0,40	-0,40	-0,55	-0,55	-0,70	-0,69	-0,85	-0,82	-0,91	-0,89	-0,86	-0,24	-0,83	-0,82
Diphenyl														
gel 60	-0,29	-0,28	-0,36	-0,38	-0,45	-0,48	-0,53	-0,53	-0,48	-0,48	-0,42	-0,36	-0,33	-0,35
gel 100	-0,34	-0,32	-0,42	-0,42	-0,55	-0,54	-0,59	-0,61	-0,59	-0,56	-0,44	-0,47	-0,49	-0,50
gel 200	-0,41	-0,41	-0,53	-0,53	-0,69	-0,69	-0,76	-0,80	-0,76	-0,78	-0,87	-0,89	-0,87	-0,87
gel 500	-0,44	-0,44	-0,58	-0,58	-0,77	-0,76	-1,00	-0,90	-1,00	-1,01	-1,02	-1,04	-0,98	-0,98
Fluoranthene														
gel 60	-0,16	-0,14	-0,28	-0,29	-0,44	-0,45	-0,50	-0,50	-0,43	-0,43	-0,34	-0,34	-0,27	-0,28
gel 100	-0,20	-0,21	-0,34	-0,35	-0,50	-0,50	-0,58	-0,58	-0,53	-0,53	-0,45	-0,45	-0,40	-0,41
gel 200	-0,32	-0,33	-0,46	-0,45	-0,65	-0,64	-0,78	-0,76	-0,76	-0,74	-0,75	-0,74	-0,72	-0,72
gel 500	-0,36	-0,36	-0,51	-0,50	-0,71	-0,72	-0,86	-0,88	-0,86	-0,85	-0,89	-0,88	-0,86	-0,87

TABLE II

Experimental and theoretical /calculated from equation 1/ R_M values fro different chromatographed substances. Mobile phase beneze-ethanol

$\varphi_{C_2H_5OH}$	0,0		0,1		0,3		0,5		0,7		0,9		1,0	
	exp	calc	exp	calc	exp	calc	exp	calc	exp	calc	exp	calc	exp	calc
Methylnaphthalene														
gel 60	-0,30	-0,31	-0,38	-0,39	-0,49	-0,50	-0,62	-0,62	-0,55	-0,56	-0,44	-0,42	-0,38	-0,39
gel 100	-0,33	-0,32	-0,43	-0,44	-0,53	-0,54	-0,65	-0,66	-0,60	-0,62	-0,51	-0,50	-0,43	-0,44
gel 200	-0,39	-0,39	-0,55	-0,54	-0,62	-0,63	-0,74	-0,74	-0,72	-0,72	-0,62	-0,62	-0,56	-0,55
gel 500	-0,41	-0,41	-0,61	-0,60	-0,66	-0,66	-0,76	-0,77	-0,78	-0,78	-0,68	-0,67	-0,60	-0,60
Chrysene														
gel 60	-0,25	-0,25	-0,32	-0,32	-0,54	-0,55	-0,59	-0,60	-0,49	-0,49	-0,40	-0,41	-0,31	-0,32
gel 100	-0,28	-0,29	-0,36	-0,37	-0,56	-0,56	-0,62	-0,62	-0,52	-0,51	-0,44	-0,44	-0,35	-0,34
gel 200	-0,38	-0,36	-0,47	-0,44	-0,61	-0,61	-0,68	-0,67	-0,59	-0,59	-0,53	-0,52	-0,46	-0,54
gel 500	-0,42	-0,42	-0,51	-0,51	-0,63	-0,63	-0,72	-0,72	-0,62	-0,62	-0,56	-0,55	-0,50	-0,50
Pyrene														
gel 60	-0,25	-0,25	-0,45	-0,46	-0,58	-0,58	-0,62	-0,62	-0,61	-0,62	-0,42	-0,41	-0,35	-0,35
gel 100	-0,29	-0,28	-0,48	-0,47	-0,62	-0,61	-0,67	-0,67	-0,66	-0,66	-0,50	-0,49	-0,44	-0,43
gel 200	-0,36	-0,36	-0,50	-0,50	-0,68	-0,69	-0,77	-0,79	-0,75	-0,76	-0,64	-0,66	-0,60	-0,61
gel 500	-0,40	-0,40	-0,53	-0,53	-0,72	-0,72	-0,85	-0,85	-0,79	-0,79	-0,72	-0,72	-0,67	-0,67
Diphenyl														
gel 60	-0,29	-0,29	-0,58	-0,41	-0,58	-0,58	-0,62	-0,62	-0,54	-0,54	-0,45	-0,45	-0,32	-0,32
gel 100	-0,33	-0,33	-0,63	-0,45	-0,63	-0,60	-0,63	-0,63	-0,59	-0,59	-0,53	-0,53	-0,40	-0,40
gel 200	-0,42	-0,41	-0,55	-0,55	-0,65	-0,66	-0,68	-0,68	-0,72	-0,72	-0,68	-0,68	-0,60	-0,60
gel 500	-0,44	-0,44	-0,67	-0,59	-0,67	-0,68	-0,71	-0,70	-0,79	-0,78	-0,72	-0,72	-0,70	-0,70

change in a regular manner. It turned out that there is a linear dependence of the R_M value on the specific surface of the adsorbent. This dependence can be presented by a linear equation.

$$R_M = as + b \quad (1)$$

where s is the specific surface area of the adsorbent, and values a and b are parameters of the straight-line.

As it appears from the graphs, the R_M values obtained from the experiment are set out on a straight-line drawn so as get the smallest deviation. Tables 3-4 shows the R_M = f (s) straight-line which were calculated and tabulated for particular concentrations of the mobile phase and the chromatographed substance. The data in Tables 3-4 indicates that the direction coefficients of straight lines grow together with the increased concentration of the more active component of the mobile phase. The differentiation of those parameters for particular substances can also observed.

The straight-lines described can be used for the calculation of the R_M values of substances on any adsorbent with a given specific surface area. A series of arduous measurements can be avoided in this way. It is enough to take two measurements of possibly extreme R_M values in order to draw straight-lines R_M = f (s). When parameters a and b of two different substances, are known the separability of such a mixture can be calculated with good approximation, with the use of the given mobile phase.

It results from Tables 1 and 2 and from the graphs of dependencies $R_{M1.2} = f(\psi_1)$ presented in Figures 6,7, based on equation 1 that those theoretical R_M values are adequate representations of real R_M values.

The obtained results seem to suggest that it is possible to tabulate values a and b in order to calculate the dependence of R_M = f (s). It has also been stated that there is a general dependence of the parameters of the straight-line R_M=as + b on the difference of the elution strength of the components of the mobile phase, and the adsorbent surface, occupied by the solute

TABLE 3

Parameters of relationship $R_M = as + b$

A. Parameters a x 10^{-3}

	Solvent system. C_6H_6-CH_3OH						
	Concentration of solvent /CH_3OH/						
Solute	0,0	0,1	0,3	0,5	0,7	0,9	1,0
Naphthalene	0,38	0,30	0,70	0,68	0,90	1,10	1,30
METHYLNAPHTHALENE	0,28	0,55	0,50	0,70	1,00	1,10	1,20
Chrysene	0,30	0,40	0,35	0,60	0,75	0,75	0,60
Pyrene	0,35	0,60	0,60	0,70	1,10	1,10	1,20
Diphenyl	0,35	0,40	0,60	0,85	1,55	1,55	1,50
Fluoranthene	0,45	0,55	0,60	0,80	1,20	1,20	1,30
	Solvent system. C_6H_6-C_2H_5OH						
METHYLNAPHTHALENE	0,25	0,50	0,40	0,32	0,55	0,55	0,50
Chrysene	0,30	0,52	0,22	0,35	0,30	0,35	0,40
Pyrene	0,35	0,12	0,31	0,50	0,38	0,70	0,68
Diphenyl	0,33	0,40	0,22	0,20	0,40	0,40	0,80

TABLE 4

Parameters of relationship R_M = as + b

B. Parameters b

Solute	0,0	0,1	0,3	0,5	0,7	0,9	1,0
	Solvent system. C_6H_6 - CH_2OH						
	Concentration of solvent /C_2H_5OH/						
Naphthalene	-0,43	-0,53	-0,65	-0,81	-0,90	-0,94	-1,12
METHYLNAPHTHALENE	-0,41	-0,53	-0,79	-0,86	-0,98	-0,99	-1,03
Chrysene	-0,46	-0,51	-0,62	-0,79	-0,73	-0,72	-0,58
Pyrene	-0,43	-0,57	-0,72	-0,85	-0,85	-0,87	-0,84
Diphenyl	-0,46	-0,59	-0,79	-0,94	-1,06	-1,05	-1,13
Fluoranthene	-0,39	-0,44	-0,64	-0,90	-0,89	-0,96	-0,92
	Solvent system. C_6H_6 - C_2H_5OH						
METHYLNAPHTHALENE	-0,33	-0,63	-0,68	-0,80	-0,71	-0,71	-0,62
Chrysene	-0,34	-0,53	-0,64	-0,73	-0,53	-0,58	-0,52
Pyrene	-0,42	-0,52	-0,74	-0,86	-0,76	-0,76	-0,72
Diphenyl	-0,47	-0,61	-0,79	-0,71	-0,75	-0,75	-0,73

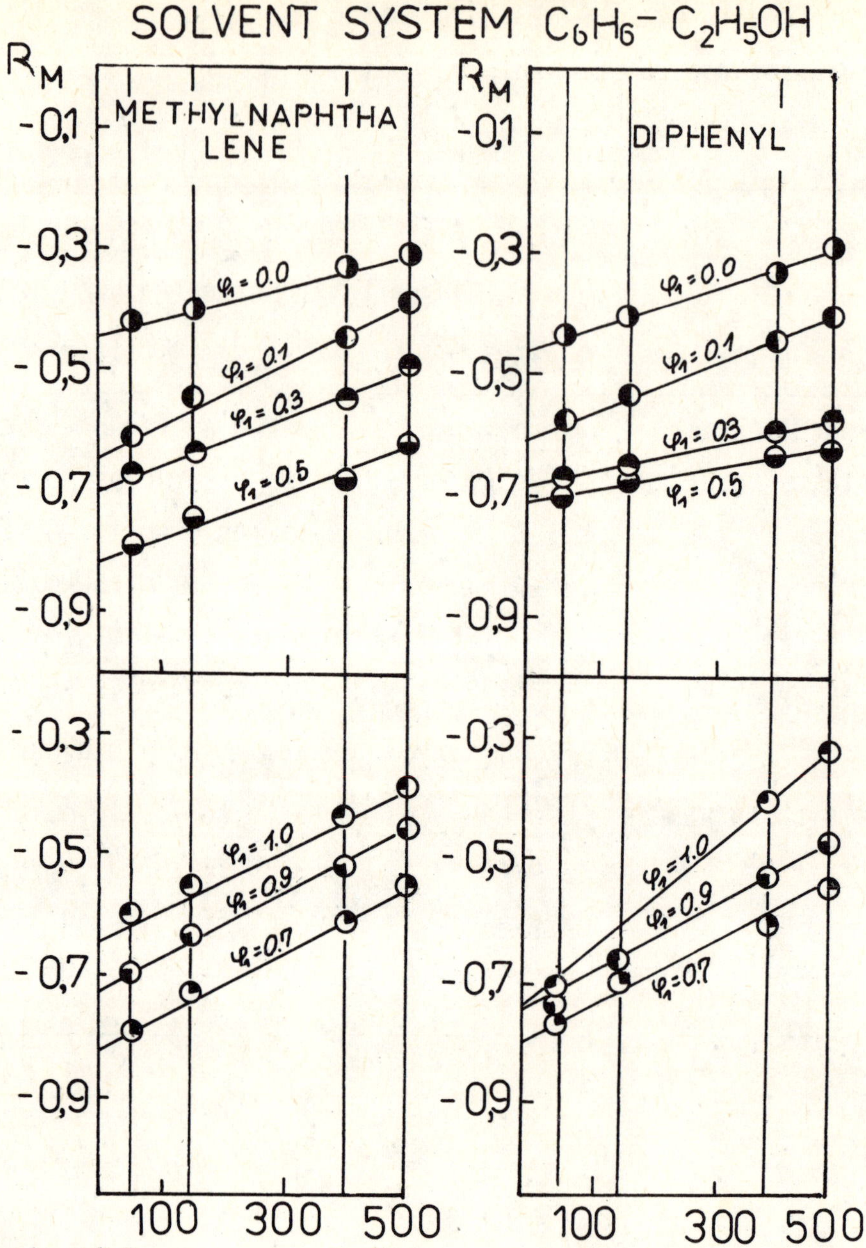

Figs 4 and 5.

Linear relationship between R_M values and specific surface area of adsorbents (mg^2/g). Solvent systems. benzene- ethanol. Experimental data, straight line - theoretical data calculated from equation 1

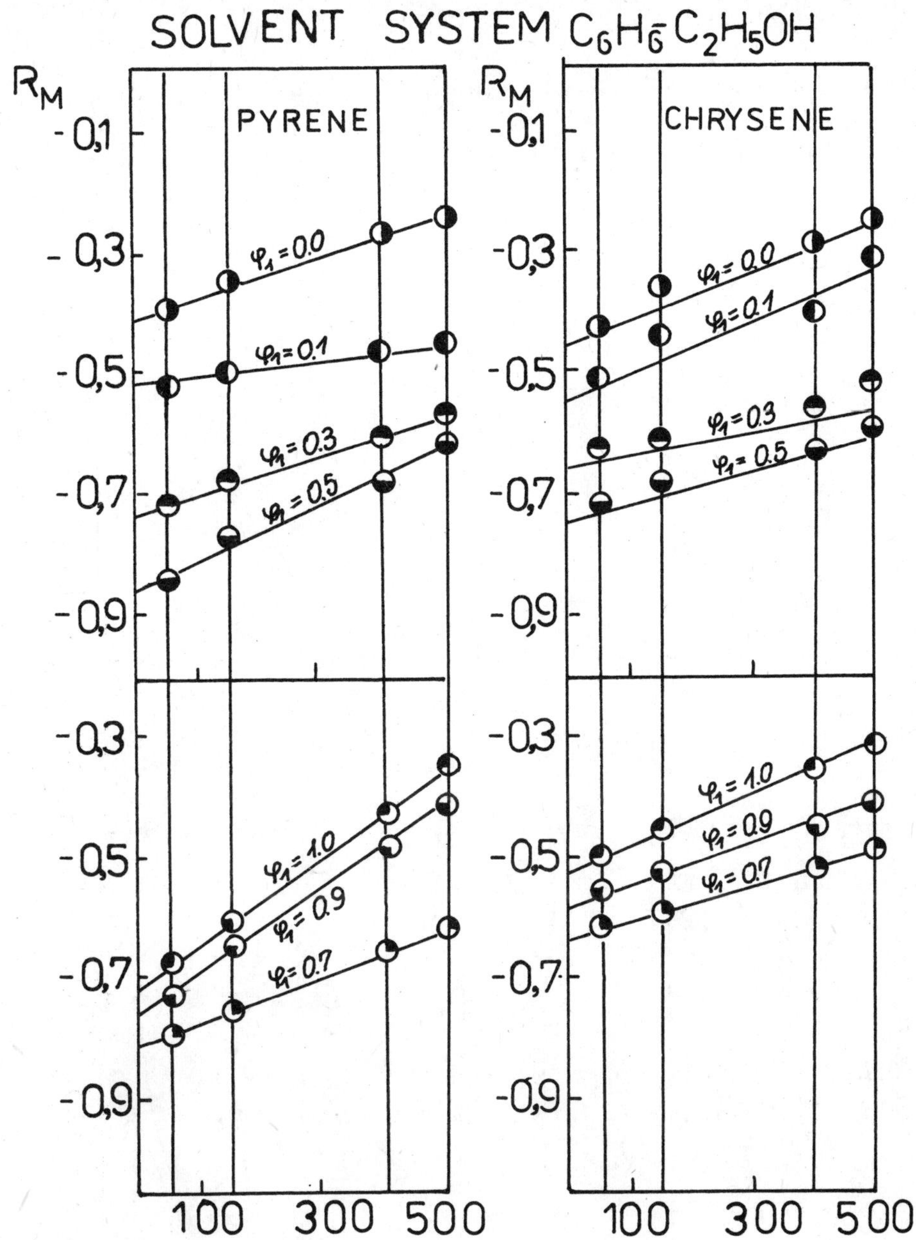
SOLVENT SYSTEM C_6H_6-C_2H_5OH
R_M
PYRENE
CHRYSENE
-0,1
-0,3
-0,5
-0,7
-0,9
φ_1 = 0.0
φ_1 = 0.1
φ_1 = 0.3
φ_1 = 0.5
φ_1 = 1.0
φ_1 = 0.9
φ_1 = 0.7
100
300
500

Fig.5

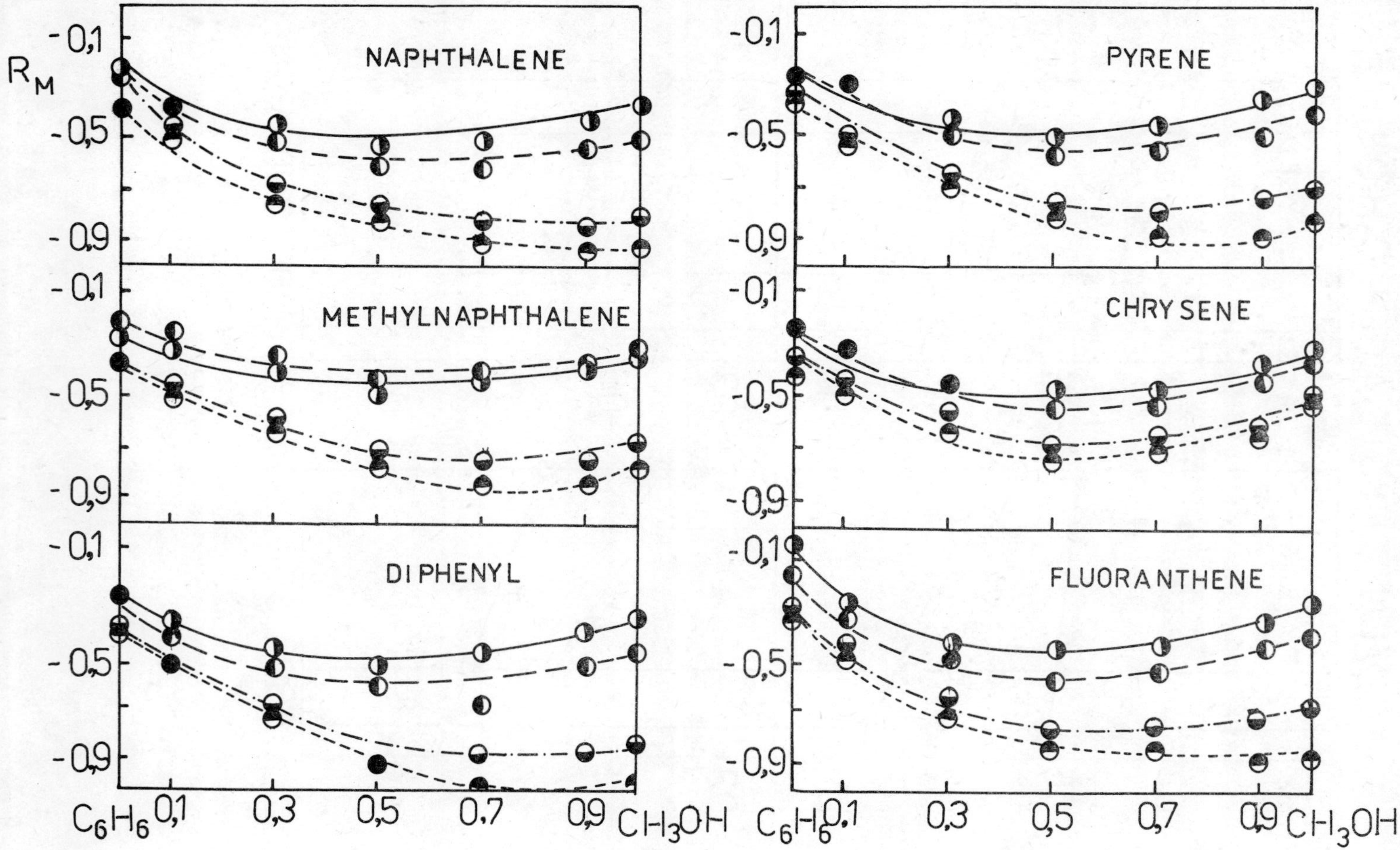

R_M
-0,1
-0,5
-0,9
NAPHTHALENE
METHYLNAPHTHALENE
DIPHENYL
PYRENE
CHRYSENE
FLUORANTHENE
C_6H_6
0,1
0,3
0,5
0,7
0,9
CH_3OH

Fig.6

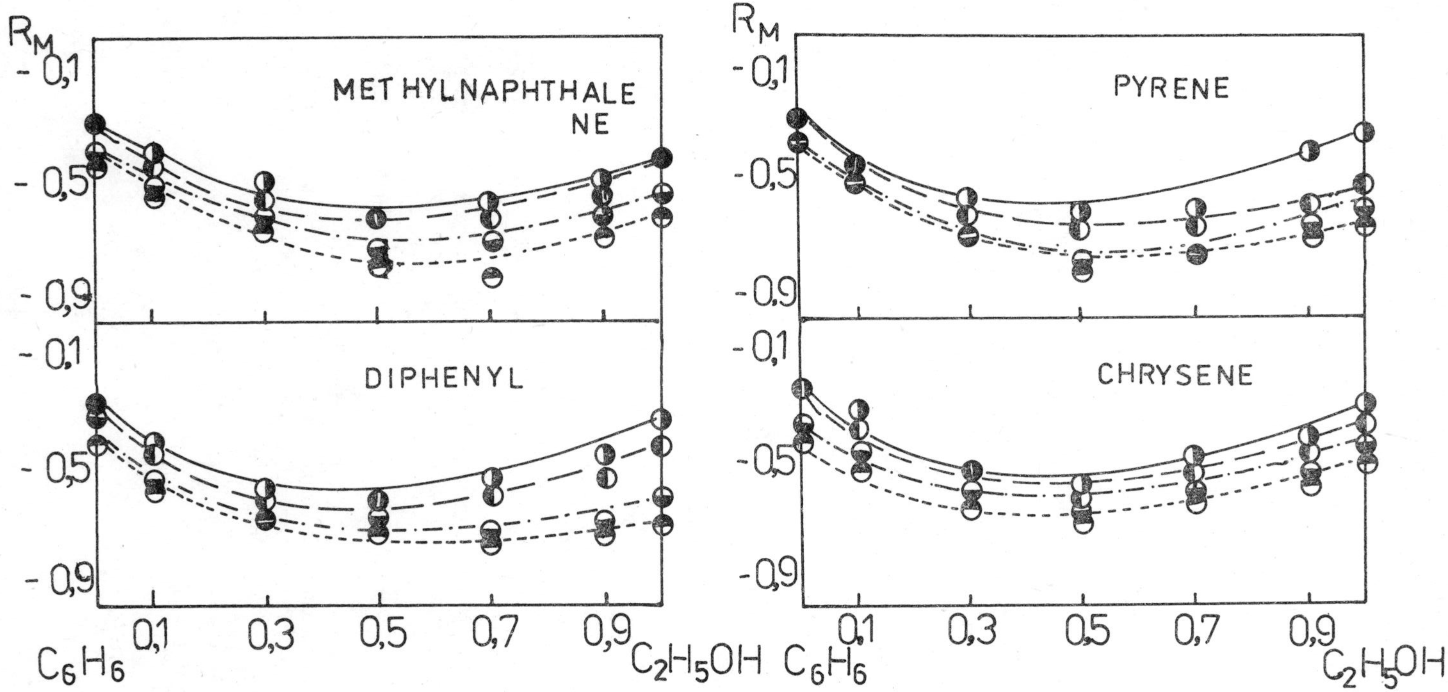

Figs. 6 and 7.

Relationships $R_M = f(\psi_1)$. Experimental data, lines-theoretical data calculated from equation 1

——— silica gel 60

— — silica gel 100

-.-.-.- silica gel 200

------- silica gel 500

and the components of the mobile phase. This affords a wider range of possibilities for optimizing the thin-layer chromatography process.

The dependencies described above can undergo changes in the case of sieve effect leading to a non-linear form of equation 1. Consequently, the applicability of the obtained dependencies should be limited to the values of the specific surface 50-500 m^2/g (14).

It seems that investigations presented above can be of great importance in transforming the chromatographic data of thin-layer chromatography to the conditions of liquid column chromatography, due to frequent differences in the properties of the stationary phase in both methods, resulting from their nature.

REFERENCES

(1) WAKSMUNDZKI,A. and ROZYLO,J. Annales UMCS, Lublin,Sec.AA, 20, 93 (1965)
(2) STAHL,E. Dünnschichtchromatograpie, Springer, I-st and II-nd EDs. 1962, 1967
(3) WAKSMUNDZKI,A. ROZYLO,J.K. Chem. Anal., 11, 101 (1968) 14, 1217, (1969), 15, 133 (1977), 16, 177 (1971)
(4) WAKSMUNDZKI,A. ROZYLO,J.K. J. Chromatog., 33, 90 (1968) 33, 96 (1968), 75, 55 (1973)
(5) HALPAAP,H. J. Chromatog., 78, 63 (1973)
(6) GEISS,F. Die Parameter der Dünnschichtchromatograpie, Vieweg, Braunschweig, 1972, p.89.
(7) SNYDER,L.R. Principles of Adsorption Chromatography,Dekker, New York, 1968
(8) ROZYLO,J.K. Interprepatation and Optimization of Adsorption Thin-Layer Chromatography Process, in Annales UMCS,Lublin 1975 /in Polish/
(9) ROZYLO,J.K. Folia Soc. Scient. Acad.Lublinensis, 21, 71 (1979)
(10) ROZYLO,J.K. J.A. Jaroniec, H. Kolodziejczyk,M.Jaroniec, J of HRC and CC, 2, 237 (1979)
(11) GETZ,M.E. Paper and Thin-Layer Chromatography Analysis of Environmental Toxicants, Heyden, London, 1980
(12) Catalog of Merck, 1981
(13) OSCIK,J. ROZYLO,J.K. Chromatographia, 4, 516 (1971)
(14) ELTEKOV,Ju.A. private communication

OPTIMIZATION POSSIBILITIES IN LIQUID COLUMN CHROMATOGRAPHY

Proc. of the Symp. on Advances in Liquid Chromatogr., Szeged, Hungary, 1982

CRITICAL EVALUATION OF OPTIMIZATION METHODS FOR HPLC

J. VAJDA, L. LEISZTNER

Institute of Forensic Science,
Budapest, Hungary

In this paper we want to discuss a few points in relation to optimization procedures for HPLC. The first thing we stress is that optimization of any process is defined only in relation to a given target, and different targets result in different optima.

Basically three approaches are possible in optimization, the first-historically also- being experimental selection of parameters for the method, less elegantly hit or miss optimization. The possibilities of this route should not be belittled, as most of the excellent separations in LC have been developed in this way, the best scheme for this was published long ago, and is part of every book and course in chromatography (1). All this only shows, that the "Organic Computer" - called the Chromatographer - is often a well programmed machine, as can be seen from the literature, but it can be indisposed, biased or just unlucky and the desired separation is not ready after wasting two months time and large amounts of LC grade solvents - so this solution of separation development is not ideal.

The second - opposite - approach could be the calculation of experimental parameters from mathematical modelling of the chromatographic processes. At present this way is not usable for diverse systems, as our knowledge of the physico-chemical processes involved in the separations is insufficient for the formulation of the mathematical functions necessary for widely applicable models. This is, we think, the method of the future and a marked increase can be seen in the number of papers reflecting this approach. The data thus amassed will serve as basis for the theory sure to come. We feel, however, that formal methods of optimization where the meaning of the symbols is as follows:

P_i	is the peak separation of the i-th pair,
P_o	is the desired minimum peak separation,
R_{s_i}	is the resolution of the i-th peak pair,
R_{s_o}	is the resolution desired,
α, β, A_i, B	are weighing factors,
t_L	is the retention time of the last peak,
t_M	is the maximum allowable separation time,
P_{inf}	is the informing power of the chromatogram, in constant time mode the information per unit time,
b	is the volume ratio of the mobile and stationary phases,
$\Delta\gamma_i$	is the difference in partition coefficient between the i-th and next peaks,
N	is the column plate number,
R	is resolution.

As can be seen from the functions above, the selection of suitable quality criterion is not satisfactorily settled.Resolution and peak separation are equally used to characterize separation. The differences are partly theoretical, but the practical aspects are also important. Using resolution means postulating Gaussian peak shapes, although we know that this is rarely the case in real chromatograms, also it is difficult to measure precisely the parameters for resolution if the overlap of the peaks is significant. Satisfying a given resolution criterion, however, means getting peak areas precise to a predefined degree, as cross contamination of peaks is below the level given by that resolution value.The use of peak separation on the other hand includes no assumption as to peak shapes, gives rapid changes in value as the separation gets better from very bad and then changes more slowly, it is easy to measure for badly separated peaks, the value changes between zero and unity only, but a given value of peak separation gives no information about the degree of cross contamination.Still the use of resolution is somewhat more common, possibly because it is easier to find manually, but authors using computers to analyse the chromatogram without human intervention usually use peak separation, as it is faster to find by computer.

The purpose functions above are limited in scope, as was discussed by Glajch (9), as even the two used in practice can not differentiate between two separations with different numbers of peaks with various resolutions. /The example quoted was four peaks with resolutions 0.4, 1.2, 1.2 and five peaks with resolutions of 0.91 for all four pairs./

From the optimization point of view samples can be classified into three types, one being a mixture of a known number of basically known compounds, the mixture of a known number of analytically significant compounds and a reasonably stable background mixture, which is also available as a sample without the important components, and thirdly a mixture of an unknown number of unknown compounds. The first two types can be handled by the CRF and COF functions described, but the third cannot.

For samples of the third type, a purpose function must be biased towards the evolution of more peaks, that is, of two chromatograms of a sample, the one with more peaks must be evaluated as better, even if the separation quality is worse. This is necessary if one wants to develop an automatic, hands off separation system for unknown samples.

We therefore propose a new purpose function, with the following form:

$$CEF = \sum_{i=1}^{n} A_i + \frac{\sum_{j-1}^{n-1} B_j P_j}{n-1} \qquad 4.$$

A_i is 1 for the i-th peak if it is analytically significant, else 0,

B_j is 1 for the j-th peak pair if at least one of the peaks of the pair is analytically significant, else 0,

P_j is the peak separation for the j-th peak pair,

n is the number of peaks in the chromatogram.

The function assigns the largest value to the best chromatogram, this is equal to n+1 in the case of baseline separation

- the middle course - has a significant role to play in developing new LC separations.

A book has recently been published by Massart et al.(2) about optimization of analytical methods. Some of these methods have already been applied to chromatography, for example the simplex technique for the optimization of the parameters of an isocratic separation of phospholipids (3) or the gradient separation of PTH-amino acids (4), the window diagram method, an application of the steepest ascent method, see the review of Laub (5), or for the optimization of the separation of weak organic (6) or aromatic acids (7). The factorial experimental design was used for the identification of narcotic drugs (8) and a method combining the last two, called overlapping resolution mapping, used in various studies (9,10). These methods all assume that the simplified model of the given system used is adequate for the solution of the problem, which assumption is verified only in a few cases.

The important question in optimization is the selection of the purpose function. Three have been published in the literature, two of them have been used in practical studies (4,8), while the third (11), is not suitable in its present form for practical applications. The functions are the Chromatographic Response Function (4) eq.1.,

$$\mathrm{CRF} = \alpha \sum_{i=1}^{n} \ln \frac{P_i}{P_o} + \beta\,(t_L + t_M) \qquad 1.$$

the Chromatographic Optimization Function (9) eq.2.,

$$\mathrm{COF} = \sum_{i=1}^{n} A_i \ln \frac{R_{s_i}}{R_{s_o}} + \beta\,(t_L - t_M) \qquad 2.$$

and the Informing power of a chromatographic run (11) eq.3.,

$$P_{inf} = 3.32\,\frac{\ln\left(1 + \frac{1}{b}\sum_{i=1}^{n} \Delta \gamma_i\right)}{\ln\left(1 + \frac{1}{\frac{\sqrt{N}}{4R} - \frac{1}{2}}\right)} \qquad 3.$$

for each peak pair. The advantage is that the chromatogram with one more peak is always evaluated as better, as the contribution of the second term is always larger than zero and not more than unity. This function can handle samples of the third category, then all constants A_i, B_j are equal to one. The function is biased in the direction of more peaks and separation is considered as of second importance only. The function can differentiate sufficiently between chromatograms with the same number of peaks, but with emphasis on average separation, so one badly separated peak pair does not cause the function value to decrease too much if the rest of the separation is good. The function does not aim at a minimum separation, but such a requirement can be incorporated at will. The time constraint included in the CRF and COF has to be left out, as it would conflict with the aim of finding the maximum number of peaks possible. A final time constraint is inherently included in the chromatography anyway, as peak broadening with increasing retention time limits peak detection after a certain time.

REFERENCES

(1) SNYDER, L.R. and KIRKLAND, J.J. /1979/ Introduction to modern liquid chromatography (2nd ed.) Wiley Interscience N.Y.
(2) MASSART, D.L. et.al. Evaluation and optimization of laboratory methods and analytical procedures, Elsevier Sci. Amsterdam, 1978.
(3) RAINEY, M.L., PURDY, W.C. /1977/ Anal.Chim. Acta 93 211-219.
(4) WATSON, M.W., CARR, P.W. /1979/ Anal.Chem. 51 1835-1842.
(5) LAUB, R.J. /1981/ Int.Lab. 11 16-29.
(6) DEMING, S.N., TUROFF, M.L.H. /1978/ Anal.Chem. 50 A546-A548.
(7) SACHOK, B. et.al. /1980/ J.Chrom. 199 317-325.
(8) LINDBERGH, W. et.al. /1981/ J.Chrom. 211 201-212.
(9) GLAJCH, J.L. et.al. /1980/ J.Chrom. 199 59-79.
(10) ISSAQ, H.J. et.al. /1981/ J.Liq.Chrom. 4 2091-2120.
(11) MASSART, D.L., SMITS, R. /1974/ Anal.Chem. 46 283-286.

Proc. of the Symp. on Advanves in Liquid Chromatogr., Szeged, Hungary, 1982

THE MOBILE PHASE IN LIQUID CHROMATOGRAPHY

H.J. ISSAQ

Chemical Carcinogenesis Program,
NCI-Frederick Cancer Research Facility,
Frederick, MD 21701, USA

INTRODUCTION

The mobile phase in GC is an inert gas (hydrogen,nitrogen,helium ...etc.) which carries the solute (in vapor form) through the liquid phase, and does not affect, or contribute to, the separation process. Retention in GC is, therefore, the result of direct interaction between the vaporized solute and the liquid phase, and separations are achieved by solute polarity differences, volatility differences, molecular shape differences or, a combination of these factors.

In contrast, the mobile phase in liquid chromatography (LC), thin layer (TLC) and high performance liquid chromatography(HPLC), plays an important, if not a major, role in the separation process. The mobile phase determines not only the separation of the components in a mixture, but the degree of resolution (how far are the peaks from each other), selectivity (the order of elution), and elution times. Mobile phase strength (polarity) determines retention time, and mobile phase composition determines selectivity and resolution.

Selection of the mobile phase in liquid chromatography is based on the properties of the stationary phase, which, in turn, is chosen after consideration of the properties of the solute mixture. This also determines the chromatographic process used, e.g. adsorption, partition or ion exchange. When pure solvents are used, they are selected from the elutropic series, in which hexane is the least, and water is the most, polar. However, when the mobile phase is a binary, or ternary mixture...the elutropic series does not apply and the following equation is used to determine the polarity of the mixed mobile phase:

$$P' = \sum_i \phi_i P_i$$

Where ϕ and P are molar fraction and polarity of solvents i in the solvent mixture which has a polarity of P'. A detailed discussion of this can be found in references 1 and 2.

The mobile phase in ion exchange is an aqueous buffer solution of known pH, which is determined by the properties of the solute (acidic or basic). The type of buffer (phosphate, acetate,etc.) its concentration (moles/liter) and pH affect the separation. The addition of organic solvent such as methanol, acetonitrile, tetrahydrofuran, etc. to the buffer solution can, in certain cases, improve the selectivity and resolving properties of the mobile phase.

The other two main chromatographic processes are adsorption and partition in which the stationary phases are silica gel and modified silica gel.

In modified silica gel, known as reversed phase, an organic moiety (R) is attached to the silynol group by silylation according to the following reaction:

$$-Si-(OH)_3 \longrightarrow -\underset{CH_3}{\overset{CH_3}{Si}}-R$$

where R is an aliphatic chain of any length ranging from C_2 to C_{22}, a phenolic group, an aliphatic chain with an $-NH_2$, -CN, or any other functional group. The addition of these groups improves the selectivity of the stationary phase due to solute-stationary phase interactions. (A review and detailed discussion of this topic is found in reference 2). The discussion here will be limited to the role of the mobile phase. Other materials can be used, but the majority of published work in both TLC and HPLC is based on the use of silica gel and reversed phases. For detailed discussion of adsorption chromatography see Reference 1.

In liquid chromatography, the relation between the solute, the mobile phase and the stationary phase is represented by the triangle shown in Figure 1. This means that there are three types of interaction; solute-mobile phase; mobile phase-stationary phase; and solute-stationary phase. Separation of a mixture is achieved when an optimum balance is reached between these three interactions. If the interaction between one pair is much stronger than with the other two, poor, or no, separation results. Also, when a mixture of solvents are used as the mobile phase solvent-solvent interaction and solvent demixing (in TLC) must also be considered.

In liquid chromatography, unlike GC, the mobile phase-solute and mobile phase-stationary phase interactions are stronger than those between solute-stationary phase, with the mobile phase-solute interactions being the most important. Examples are given later.

In partition chromatography separations are a function of the partition coefficient of the solutes between the mobile and stationary phases. The reader is referred to the work of Scott (3) on the bilayer theory. If all the solutes in the mixture have the same partition coefficient, then the solute mixture elutes as one peak, i.e., no separation. However, if the partition coefficients of all the components of the mixture are different, separation of all the components result. Thus, if the solute mixture favors one phase more than another and resides in that phase, poor, or no, separation is obtained. However, if the solutes reside in both phases, and these retention times are different for each component resolution is obtained. In other words, to resolve the components of a mixture the chromatographer must create conditions under which the solutes are forced each to favor both mobile and stationary phases, differently. To achieve this either the stationary phase, the mobile phase, or both can be altered. If separation cannot be achieved on silica gel due to the polarity of the solutes, a nonpolar stationary phase must be selected. The analyst can also change the mobile phase from hexane to ethyl acetate, methanol, or any organic solvent selected from the elutropic series when silica gel or alumina is the stationary phase. Or an alternative is to change from methanol/water to acetonitrile/water, or to change the ratio of water/organic modifier when a reversed phase column or plate is being used. If, when the solute mixture is spotted on the TLC plate or injected into the column, it does not elute, the mobile phase used can be changed since it is clear that the solute mixture prefers the stationary phase (strong solute-stationary phase interaction but weak mobile phase-solute interaction). Alternatively, the stationary phase can be changed when the solute-stationary phase interaction is weaker. However, if the solute moves with the solvent front the reverse is true, i.e., strong solute-mobile phase or weak solute-stationary phase interactions. Then either the mobile phase or stationary phase must be changed. It is, of course, easier and cheaper, especially in HPLC, to change or modify the mobile phase. When pure organic solvents are used as the mobile phase, the elutropic series gives an excellent guide to the selection of an appropriate mobile phase depending on the properties of the solutes. When a mixture of organic or organic/aqueous solvents is used, the task of selecting an optimum mobile phase becomes more complex (4-7) and prediction of the retention times more difficult (4). The analyst would, in such cases, need a statistical technique to obtain resolution in the minimum amount of time. Recently, new approaches have been published (8-12) these will be reviewed later. These methods of isocratic mobile phase selection for optimal separation are not only sound, but can save the analyst time and materials, and make the analysis more cost effective.

EXAMPLES AND DISCUSSION

The first topic to be discussed is solvent strength or polarity because it has been assumed that solutes chromatographed in mobile phases with the same solvent strength will have the same retention times. The relation between retention time and solvent strength is described by the following equation:

$$\log k_1'/k_2' = \alpha A_s(\varepsilon_2^0 - \varepsilon_1^0) \qquad \text{(I)}$$

Where k' is the capacity factor, and ε^0 is the solvent strength parameter of solvents 1 and 2. A_s is the molecular area of the adsorbed sample and α is adsorbent surface activity function,

$$K' = (R_t - R_{to}) / R_{to} \qquad \text{(II)}$$

Where R_{to} is the retention time of unretained solute. It was found that the relation in equation I does not always hold (4, 13,14). For example, when the mobile phases (acetonitrile/water and methanol/water) which have the same solvent strength calculated according to eq. I were used to elute the same solutes /naphthalene and biphenyl) on the same solid phase (C_{18}) the retention times were not the same (4). A term was later added to eq. I which accounted for solvent-solute interactions (13,14).

Since different solvents give different selectivities (2, 4,15-18), changing the solvent composition may result in different elution orders depending on the properties of the sample mixture and the solvent chosen. For a mobile phase mixture, in general solvent strength (polarity) determines the elution distance of the solutes (i.e. R_t) when mixtures of the same binary solvents are used, while mobile phase composition determines the selectivity. The composition of the mobile phase also determines the degree of separation (α), between two adjacent peaks i and ii, where

$$\varepsilon = k'_{ii}/k'_{i} \qquad \text{(III)}$$

Based on Snyder's theory (12), Saunders (17) presented a graphical representation based on ε for selecting a solvent for adsorption liquid chromatography. The application of these graphs is rapid and provides a reasonable first approximation to a solvent mixture appropriate for a given sample. It must be stressed that the results are only approximate; in some cases the solvent mixture will not be ideal.

For a given sample and adsorbent, log K' varies linearly with ε. This is generally true for k' values between 1 and 10, which is an acceptable working range allowing separation of a component from the mixture, without leading to dilution of the sample, or long retention times. Solvent strength (polarity) gives a general indication of solute retention but it may not predict the correct retention times (4,8,16,18). Figures 2 and 3 and Table I illustrate this point.

TABLE I

EFFECT OF BINARY SOLVENT COMPOSITION ON R_f AND α VALUES IN REVERSED PHASE TLC (P'= 4.3)

SOLVENT	R_f X100 BeP	R_f X100 ANTHRACENE	α
ethanol	63	73	1.16
n-propanol/water	60	70	1.17
n-propanol/methanol	71	81	1.14
n-butanol/water	60	76	1.27
n-butanol/methanol	72	78	1.08
i-propanol/methanol	56	68	1.21
t-butanol/methanol	49	60	1.24
n-butanol/acetone	80	85	1.06
n-propanol/acetone	75	81	1.08

For other examples see reference 4.

Another aspect of the role of the solvent in liquid chromatography is the selectivity of the mobile phase chosen. Solutes may not elute in the same way in different mobile phases; a good example is given in Table II:

TABLE II

EFFECT OF MOBILE PHASE COMPOSITION ON SOLUTE ELUTION ORDER

MOBILE PHASE	ORDER OF ELUTION*
72% Methanol/water	N, A, B, M, E
64% Actontrile/water	A, N, M, B, E
42% THF/water	A, M, N, E, B

Where N = naphthalene, B = biphenyl, A = anthraquinone, M = methylanthraquinone, E = ethylanthraquinone

*C_{18} reversed phase column was used.

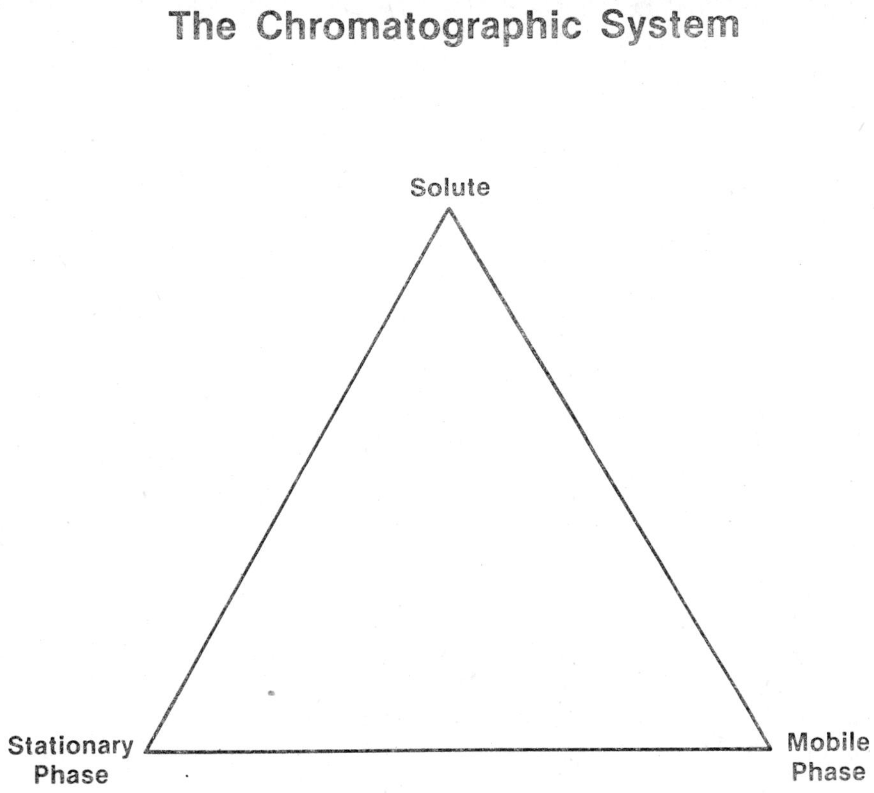

Figure 1. The Chromatographic process

Another aspect of reversed phase chromatography is that the volume of water affects the resolution and retention times of a solute mixture. The more organic in nature the solute modifer the larger is the percentage of water in the binary mobile phase. The percentage of water in methanol/water, acetonitrile/water and THF/water increases from say 20% to 35% to 55% respectively. This may be due to the fact that organic modifiers, which would dissolve the organic solutes, require water to achieve different solute distribution coefficients. Water also aids in the formation of a bilayer on the bonded alkyl chain (3) which affects the distribution of the solute between the mobile and stationary phases which affects in the separation.

Table III shows how water influences the separation on a C_{18} reversed phase column when different aliphatic alcohols are used as organic modifiers.

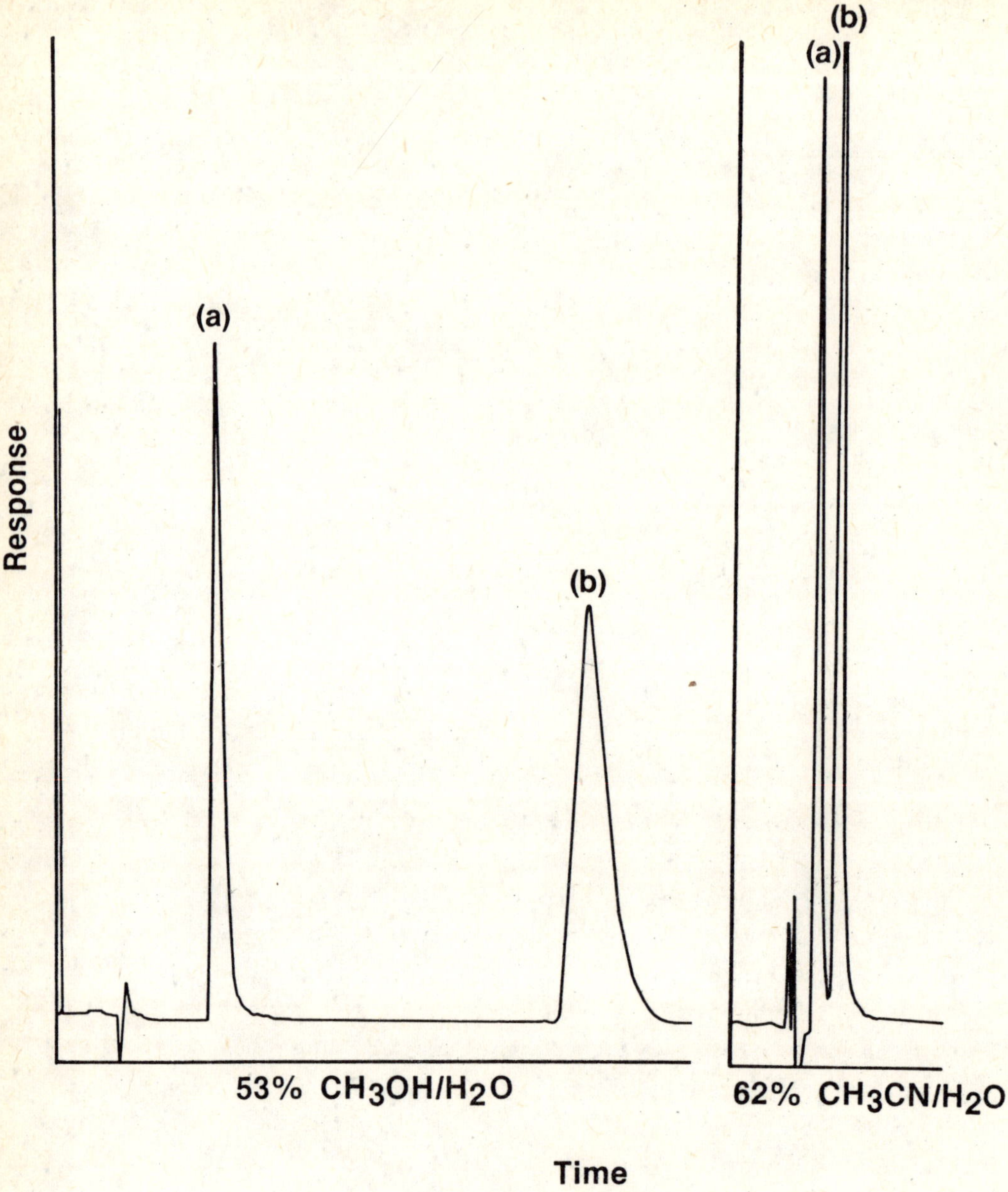

Figure 2. Comparative HPLC separation of dimethylphthalate (a) and diethyl-phthalate using a reversed phase C_{18} column in two different mobile phases having the same polarity; 53% methanol/water and 62% acetonitrile/water at a flow rate of 1.2 ml/min.

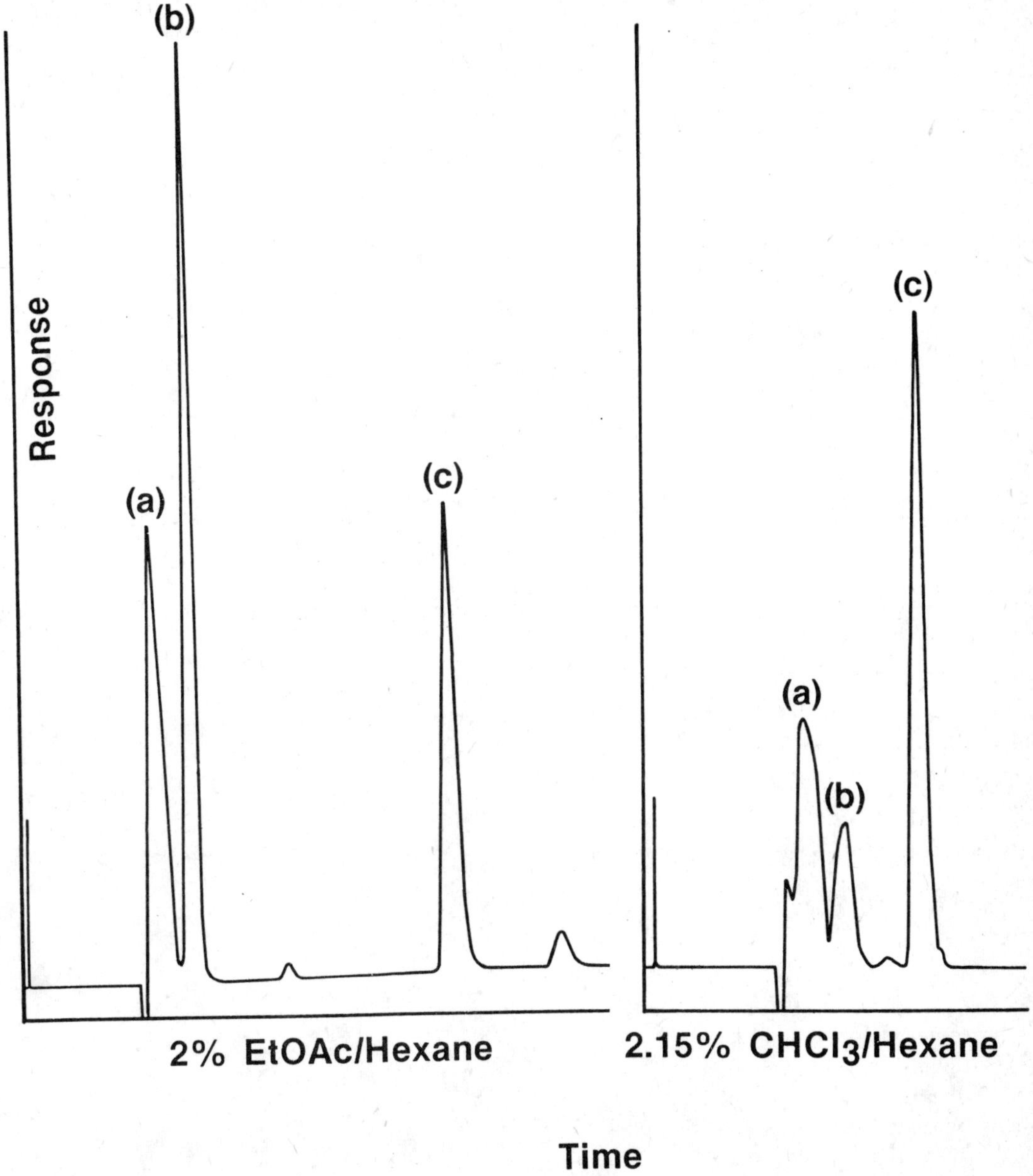

Figure 3. Comparative HPLC separation of benzo [a] pyrene (a), naphthacene (b) and anthracene (c) using a μ Porasil column and two different mobile phases having the same polarity; 2% ethyl acetate/hexane (left) and 2.15% chloroform/hexane (right)

TABLE III

EFFECT OF WATER CONTENT ON THE SEPARATION OF BENZO(e)PYRENE AND ANTHRACENE ON C_{18} REVERSED PHASE PLATES, USING ALCOHOLS OF DIFFERENT NORMAL CHAIN LENGTH

ALCOHOL	%	α
Methanol	95	1.74
Ethanol	100	1.16
	85	1.52
	78	1.65
n-Propanol	95	1.17
	83	1.34
	74	1.46

The next example, Table IV, shows that when the right mobile phase is used separation is achieved. The table shows that the separation of aflatoxins B_1, B_2, G_1 and G_2 on Backerflex silica gel plates was achieved in chloroform:acetone:ammonium hydroxide (90:10:0.25) and in chloroform:acetone:hexane (85:15:20) but not in the other ten mobile phases.

TABLE IV

EFFECT OF MOBILE PHASE ON THE SEPARATION OF AFLATOXINS B_1, B_2, G_1 AND G_2 ON BAKERFLEX SILICA GEL PLATES

SOLVENT	RATIO	$R_f \times 100$ B_1	B_2	G_1	G_2
Chloroform:Acetone:Water	88:12:1.5	63	63	52	52
Chloroform:Tetrahydrafuran	90:10	48	48	33	33
Chloroform:Acetone	90:10	42	37	37	31
Benzene:Methanol:Acetic Acid	90:5:5	27	27	20	20
Chloroform:Acetic Acid:Ether	17:1:3	99	99	88	88
Chloroform:Acetone:Ammonium Hydroxide	90:10:0.25	63	57	52	46
Chloroform:Acetone:Hexane	85:15:20	68	61	52	45
Chloroform:Methanol	4:1	96	96	96	96
Toluene:Ethyl Acetate:Formic Acid	6:3:1	99	76	76	61
Toluene:Ethyl Acetate:Chloroform:Formic Acid	70:50:50:20	99	89	89	76
Benzene:Ethanol	95:5	95	95	82	82
Chloroform:Methyl Isobutyl Ketone	4:1	99	95	45	76

Bakerflex silica gel plates, 0.1 mm layer.

The solvents in Table IV were selected mainly by trial and error based on the properties of both the solute and the stationary phase. Systematic approaches to the selection of a mobile phase that will give optimum separation using normal and reversed phase TLC plates (9) and HPLC columns (8-12) have been published. The methods are based on statistical calculations, whereby peak pair resolution is plotted versus mobile phase composition. The resulting overlapping resolution mapping (ORM) plots indicate areas of maximum pair resolution. The union of the peak pair(ORM)plots will give the region where all resolutions are above a level predetermined by the analyst. The procedure followed is very simple and is summarized below for the methods of Glajch et al (8) and Issaq *et al* (9).

Combinations of three solvents A, B and C are devised according to Table V; other combinations may also be used. The initial solvents may be pure or a mixture of solvents, two organic for normal phase, water/organic modifier for reversed phase. After selecting the solvents and proportions (Table V), ten run one for each solvent combination are made. These are used to calculate the resolutions of each pair of compounds in the mixture. If no peak crossover takes place the resolution between each pair (1-2, 2-3, 3-4...etc) is used. If peak crossover does occur the resolution between all the peaks is calculated (1-2, 1-3, 1-3,1-4, 2-3, 2-4, 3-4...etc.), and used in determining the optimum mobile phase.

In the Issaq *et al* system (9), two computer programs are used to predict optimum solvent composition. The first is a FORTRAN program (PEAKIN) which rearranges resolutions to correct for crossover and produces a data file suitable for use in the next program. The second program is a SAS (Statistical Analysis System - version 79.5) route (19). A DATA paragraph converts the three-dimensional solvent compositions to a two-dimensional triangle representation as used by Snee (20). The data are fitted into a cubic model for a three-dimensional system. The parameters of the cubic equation for each set of peak resolutions are computed using the general linear model (GLM) procedure. The PRINT procedure lists predicted resolutions of each peak pair for all solvent combinations varying each solvent from zero to 100 percent by

2% increments (Table VI). Contour plots, of the region where the predicted resolution above a desired level determined by the analyst, are produced using the PLOT procedure. The union of these plots showing the region where all resolutions are above this level, and plots showing the area of maximum total resolution, are also produced using PLOT. A flow chart of the procedure is shown in Fig.4. (See Ref. 9 for details.)

Ideally, where a combination of three modifiers and a base solvent is used the region of the optimum mobile phase mixture found from the ORM calculations will be in the center of the triangle which means that all the solvents used contribute to the separation process. If one of the modifiers (A) is not ideal, the optimum mixture will be composed of the other two modifiers (B and C), with only a small amount of A. Therefore, the optimum region can indicate which of the three modifiers is a poor choice. The base solvents are water for reversed phase, and hexane (8) or chloroform (9) for normal phase. Other solvents for normal phase may also be used. A good correlation between predicted and experimental resolution was found (9).

TABLE V

COMBINATION OF SOLVENTS A, B AND C USED IN THIS STUDY TO PREDICT OPTIMUM MOBILE PHASE COMPOSITIONS

% Solvent A	% Solvent B	% Solvent C
100	0	0
0	100	0
0	0	100
50	50	0
50	0	50
0	50	50
33	33	33
67	16	16
16	67	16
16	16	67

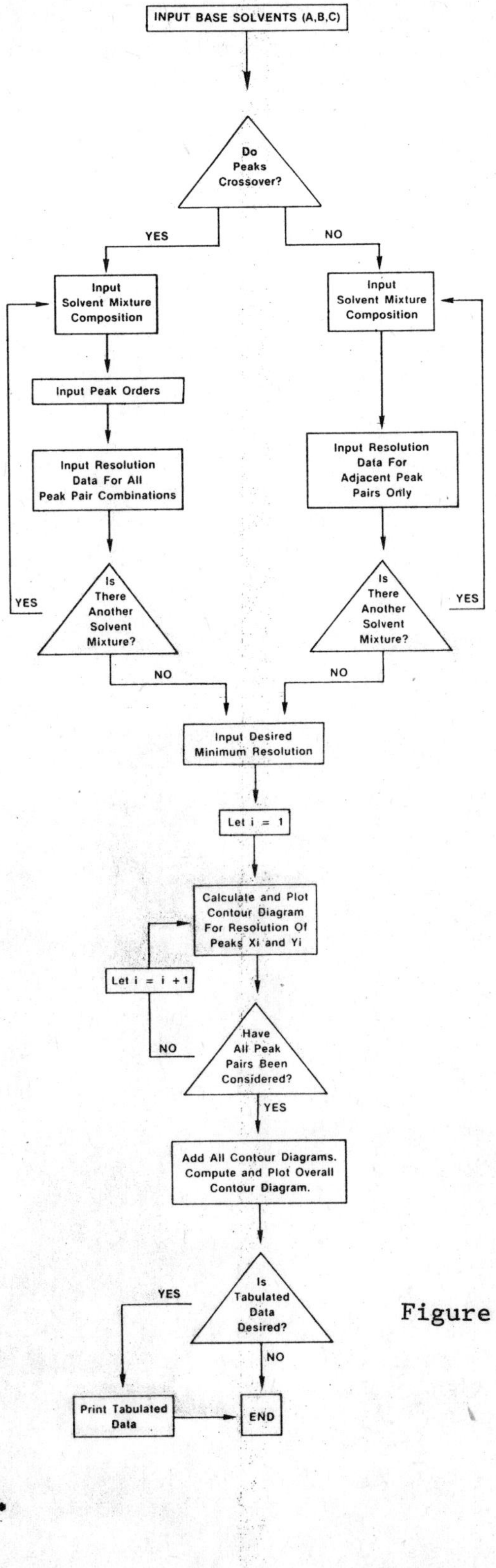

Figure 4 Flow chart of the procedure used in the solvent optimization study

TABLE VI

TABULATION OF RESOLUTION VALUES AND MOBILE PHASE COMPOSITION AS PREDICTED BY THE COMPUTER

S T A T I S T I C A L   A N A L Y S I S   S Y S T E M

GTRES=1 - RESOLUTION>2.25   GTRES=0 - RESOLUTION<=2.25

SOLVENT STUDY - A=CH3OH,B=CH3CN,C=THF

OBS	A	B	C	PEAK1	PEAK2	RESPRED
11761	0.66	0.26	0.08	1	2	1.2342
11762	0.66	0.26	0.08	1	3	6.3273
11763	0.66	0.26	0.08	1	4	8.4557
11764	0.66	0.26	0.08	1	5	10.3659
11765	0.66	0.26	0.08	2	3	5.0931
11766	0.66	0.26	0.08	2	4	7.2215
11767	0.66	0.26	0.08	2	5	9.1306
11768	0.66	0.26	0.08	3	4	2.1283
11769	0.66	0.26	0.08	3	5	4.0378
11770	0.66	0.26	0.08	4	5	1.9091
11771	0.66	0.28	0.06	1	2	0.9280
11772	0.66	0.28	0.06	1	3	6.5180
11773	0.66	0.28	0.06	1	4	8.5569
11774	0.66	0.28	0.06	1	5	10.7119
11775	0.66	0.28	0.06	2	3	5.5900
11776	0.66	0.28	0.06	2	4	7.6290
11777	0.66	0.28	0.06	2	5	9.7842
11778	0.66	0.28	0.06	3	4	2.0389
11779	0.66	0.28	0.06	3	5	4.1941
11780	0.66	0.28	0.06	4	5	2.1552
11781	0.66	0.30	0.04	1	2	0.6061
11782	0.66	0.30	0.04	1	3	6.7223
11783	0.66	0.30	0.04	1	4	8.6888
11784	0.66	0.30	0.04	1	5	11.1004
11785	0.66	0.30	0.04	2	3	6.1163
11786	0.66	0.30	0.04	2	4	8.0827
11787	0.66	0.30	0.04	2	5	10.4960
11788	0.66	0.30	0.04	3	4	1.9665
11789	0.66	0.30	0.04	3	5	4.3792
11790	0.66	0.30	0.04	4	5	2.4133
11791	0.66	0.32	0.02	1	2	0.2685
11792	0.66	0.32	0.02	1	3	6.9403
11793	0.66	0.32	0.02	1	4	8.8513
11794	0.66	0.32	0.02	1	5	11.5313
11795	0.66	0.32	0.02	2	3	6.6718
11796	0.66	0.32	0.02	2	4	8.5828
11797	0.66	0.32	0.02	2	5	11.2661
11798	0.66	0.32	0.02	3	4	1.9110
11799	0.66	0.32	0.02	3	5	4.5933
11800	0.66	0.32	0.02	4	5	2.6833
11801	0.66	0.34	0.00	1	2	-0.0847
11802	0.66	0.34	0.00	1	3	7.1718
11803	0.66	0.34	0.00	1	4	9.0443
11804	0.66	0.34	0.00	1	5	12.0048
11805	0.66	0.34	0.00	2	3	7.2565
11806	0.66	0.34	0.00	2	4	9.1291
11807	0.66	0.34	0.00	2	5	12.0943
11808	0.66	0.34	0.00	3	4	1.8725
11809	0.66	0.34	0.00	3	5	4.8363

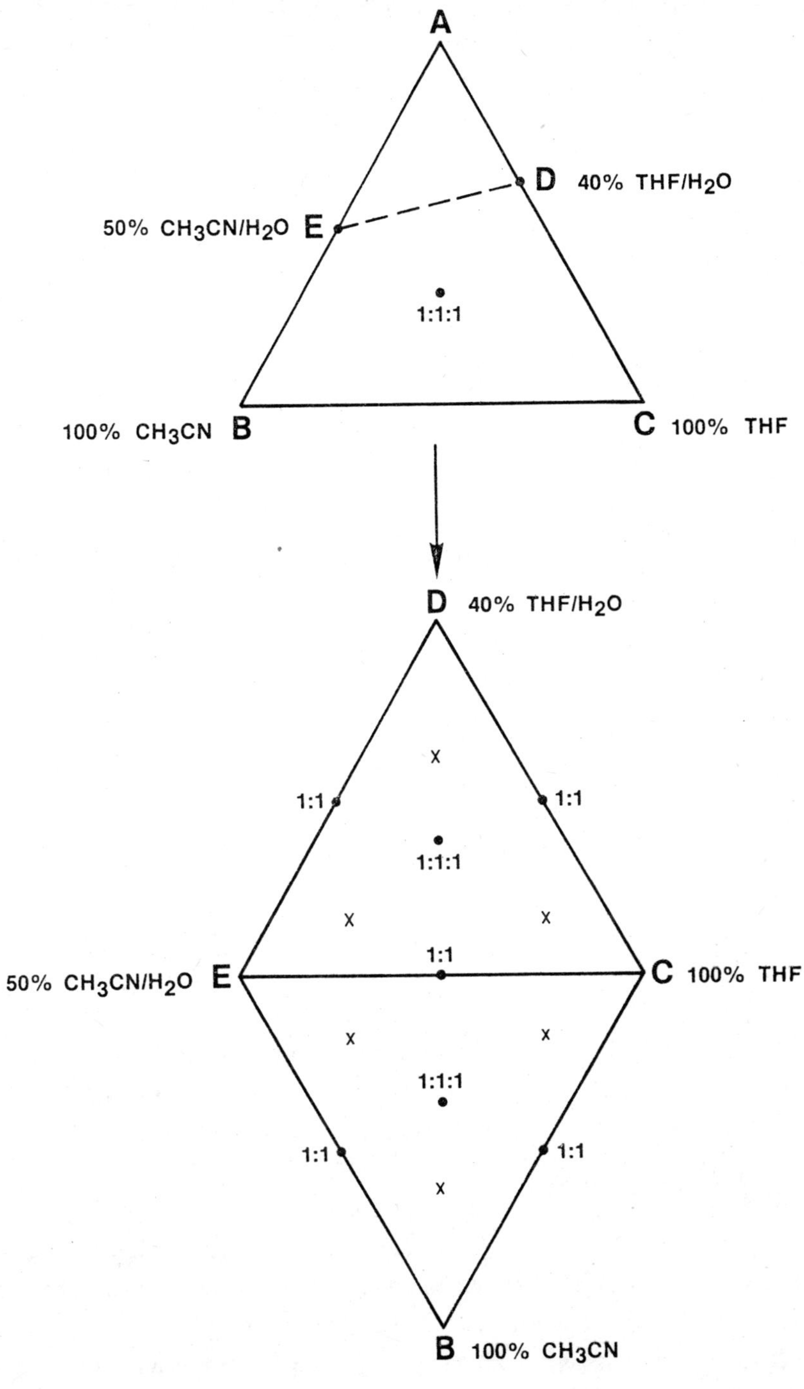

Figure 5 Solvent selection using two organic modifiers according to Belinsky. See text for details.

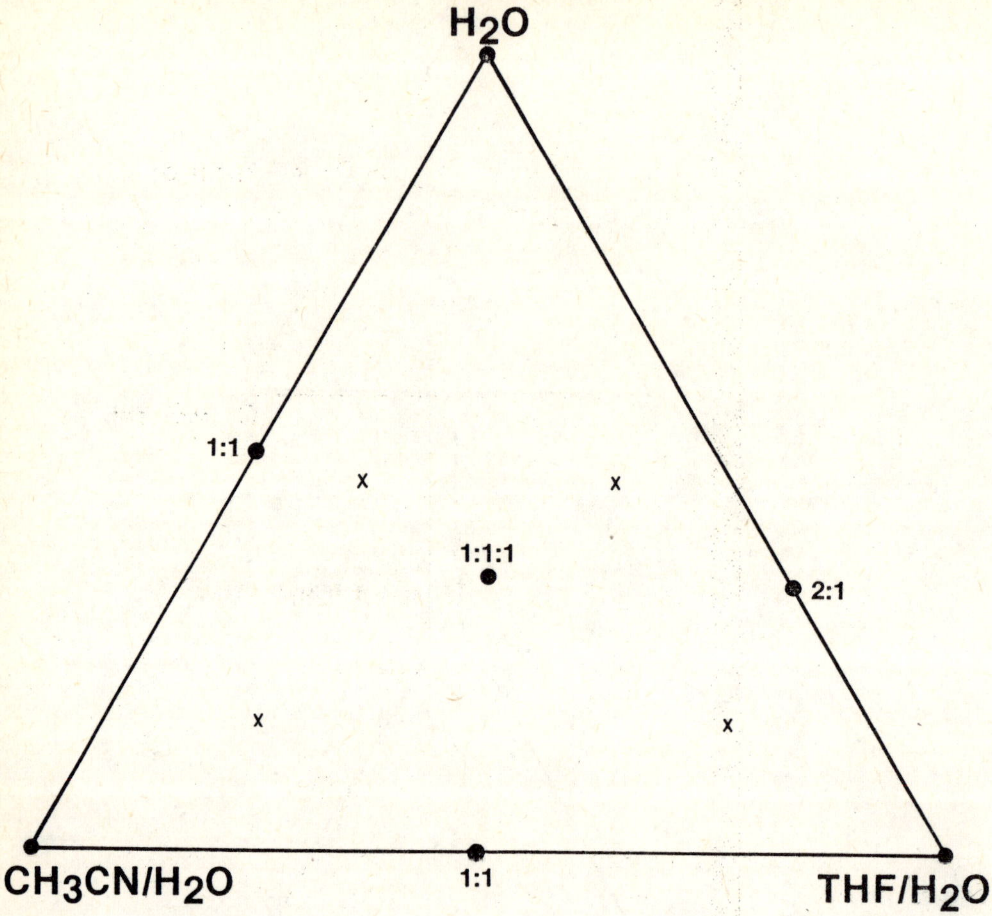

Figure 6 Solvent selection using two organic modifiers as described in the present study

Belinky (21) used two organic modifiers and water to achieve separations in reversed phase HPLC. His system described in Fig. (5) required 17 data points to achieve an optimum mobile phase.

Our approach to two organic modifiers is simpler and requires only ten data points (see Fig.6). A computer program (9) is used to select the mobile phase which will give optimum resolution of the components in a mixture.

The contour plot for two organic modifiers is shown in Fig. (7). Table VII compares predicted and experimental peak pair resolutions using 95% methanol/water and 75% 2-ethoxyethanol/ water.

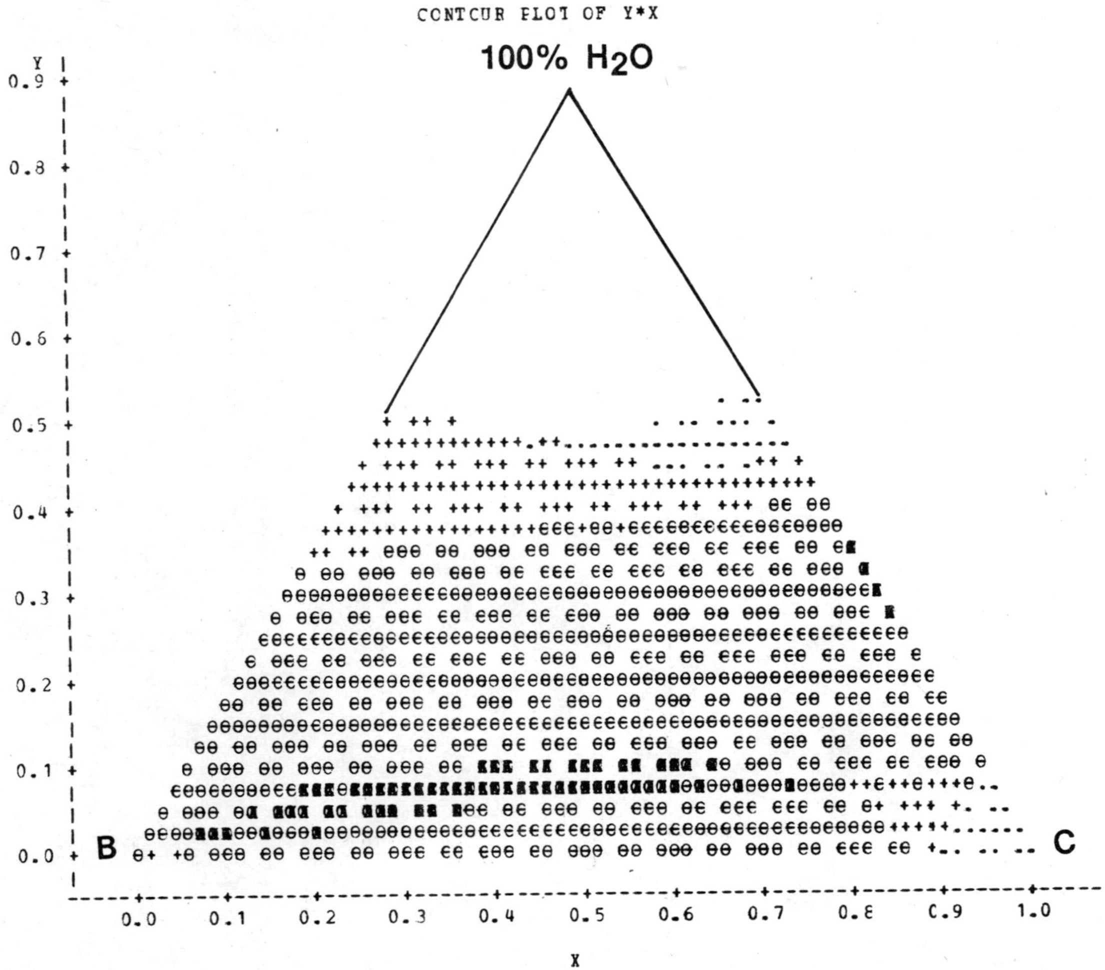

Figure 7 Statistical solvent selection using two organic modifiers, B and C

TABLE VII

PREDICTED AND EXPERIMENTAL PEAK PAIR RESOLUTIONS FOR N-M, M-E AND E-D, USING REVERSED PHASE C_{18} TLC PLATES AND SELECTED MOBILE PHASES FROM THE CONTOUR PLOT

SELECTED MOBILE PHASE	COMPOUND	PREDICTED RESOLUTION	EXPERIMENTAL RESOLUTION
95% Methanol: 75% 2-Ethoxyethanol	D		
(2:3)	E	8.1	8.1
	M	5.8	5.4
	N	3.5	3.6
95% Methanol: 75% 2-Ethoxyethanol	D		
(3:2)	E	7.9	7.9
	M	5.1	5.0
	N	3.5	4.2

Where: N = naphthalene   E = 1-ethylnaphthalene
M = methylnaphthalene   D = 1,3-dimethylnaphthalene

Another approach to solvent optimization is the window diagram technique developed by Laub and Purnell (11) and based on a mathematical functional relationship between resolution and mobile phase composition. The original paper should be consulted for details. This technique has been used to separate a nine component mixture (12).

Recently (22), another approach to solvent optimization was published based on the linear relationship between log K' and log mole fraction of the solvent. This approach is not as sound or general as the statistical approaches discussed earlier.

CONCLUSION

It is clear from the discussion above that the mobile phase plays an important role in liquid chromatography. Recently, selection of the mobile phase for optimal separation has been simplified, and it is no longer merely a function of operator intuition, or a trial and error approach, but is based on sound scientific techniques.

Mobile phases are selected on the basis of the properties of the sample and the stationary phase which, in turn, determine the chromatographic process chosen. Different selectivities and elution times may be obtained by changing the mobile phase nature and composition.

In conclusion, subtle alterations of the mobile phase give the liquid chromatographer a powerful tool for separating complex mixtures.

REFERENCES

(1) SNYDER, L.R: (1968) Principles of Adsorption Chromatography, Marcel-Dekker, Inc. NY.NY.

(2) SNYDER, L.R. and KIRKLAND, J.J.: (1979) Introduction to Modern Liquid Chromatography, Wiley, NY. 2nd Ed.

(3) SCOTT, R.P.W. and KUCERA, P.: (1979) J.Chromatogr. 171 37

(4) ISSAQ, H.J., KLOSE, J.R. and CUTCHIN, W.: (1982) J.Liquid Chromatogr. 5 652

(5) SCHOENMAKERS, A.J., BILLET, H.A. and DeGALAN, L.: (1981) J.Chromatogr. 218 261

(6) JARONIEC, M. and JARONIEC, J.A.: (1981) J.Liquid Chromatogr. 4 2121

(7) SOCZEWINSKI, E.: (1980) J. Liquid Chromatogr. 3 1781

(8) GLAJCH, J.L., KIRKLAND, J.J. and SQUIRE, K.M.: (1980) J.Chromatogr. 199 57

(9) ISSAQ, H.J., KLOSE, J.R., McNITT, K.L., HAKY, J.E. and MUSCHIK, G.M.: (1981) J.Liquid Chromatogr. 4 2091

(10) GLAJCH, J.L., KIRKLAND, J.J., and SNYDER, L.R.: (1982) J. Chromatogr. 238 269

(11) LAUB, R.J., and PURNELL, J.H.: (1975) J.Chromatogr. 112 71

(12) SACHOK, R., KONG, R.C., and DEMING, S.N.: (1980) J.Chromatogr. 199 317

(13) JARONIEC, M., ROZZYLO, J.K., KARONIEC, J.A., and OXCIK-MENDYK, B.: (1980) J.Chromatogr. 188 27

(14) MARTIRE, D.E., and BOEHM, R.E.: (1980) J.Liquid Chromatogr. 3 753

(15) SNYDER, L.R.: (1974) J.Chromatogr. Sci. 92 223

(16) BAKLAYAR, S.R., McIWRCK, R., and ROGGENDORF, E. (1977) J.Chromatogr. 142 343

(17) SAUNDERS, D.L.: (1974) Anal.Chem. 46 470
(18) ROGERS, D.: Am.Lab.May, 1980, P.49.
(19) SAS Institute, SAS User's Guide, 1979 Edition.
(20) SNEE, R.D.: (1979) Chem.Tech. 9 702
(21) BELINKY, B.R.: Analytical Technology and Occupational Health Chemistry, ACS Symposium Series, Volumn 220 pp 149-168, American Chemical Society, Washington, D.C., 1980
(22) HARA, S., KUNIHIRO, K., YAMAGUCHI, H., and SOCZEWINSKI, E.: (1982) J.Chromatogr. 239 687

ACKNOWLEDGEMENTS

"By acceptance of this article, the publisher or recipient acknowledges the right of the U.S. Government to retain a nonexclusive, royalty-free license in and to any copyright covering the article."

Proc. of the Symp. on Advances in Liquid Chromatogr., Szeged, Hungary, 1982

OPTIMIZATION OF GEL CHROMATOGRAPHIC SEPARATIONS

Numerical evaluation of gel chromatographic elution curves, optimal sample size and fractionation

H. KALÁSZ, J. NAGY, L. KERECSEN

Department of Pharmacology, Semmelweis University of Medicine, Budapest, Hungary

SUMMARY

Points of gel chromatographic elution curves were approximated with the aid of the Gaussian formula. The calculated values were obtained in reverse transmission (100-T%) units. The superposition of several values can be performed after the (100-T%)units had been transformed into extinctions to obtain the points of the chromatographic elution profile, the extinction values have to be transformed back into reverse transmission. A single elution curve as well as the superposition of several such curves may be calculated with a Hewlett-Packard HP-97 calculator using simple programs.

The problem of optimal sample size is analyzed and the effect of increasing sample size is investigated with the aid of calculated curves.

The fractionation of the effluent according to the standard deviation of the elution curve has been suggested. A formula which is easy to calculate is introduced for the estimation of the number of fractions to be collected until the peak maximum, namely

$$n_{fr} = \sqrt{N}$$

that is the number of fractions is equal to the square root of the theoretical plates.

The correctness of this statement, namely that it means an optimal fractionation, is verified by the help of a calculation program and different calculations.

Calculation program was composed for the simulation of the joint effect of increasing sample size and collection of fractions with different sizes.

INTRODUCTION

The numerical calculation of elution curves is widely used both in gas and in liquid column chromatography. The asymmetry of peaks however requires the application of rather complicated mathematical formulas.

Buys, de Clerk and Pretorius (1-5) elaborated the method of

equivalent Gaussian inlet profile as well as the first, second and third moment analysis of non-linear chromatography. Both the mathematical formulas and the methods are really brilliant but somehow too complicated for the practical use of everyday work.

The gel chromatographic elution curves can also be approximated by numerical calculations. The process and mathematical models were discussed in detail during the Symposium on Analytical Gel Permeation Chromatography (6). Pierce and Armonas (7) described their method of solving Tung's Axial Diffusion Equation (8) for gel permeation chromatography. Meyerhoff (9) introduced the comparison of some gel permeation elution curves and the calculated ones. Pickett et al (10) discussed the problem of reshaping of elution curves.

Earlier data on the sample size parameter reported that the maximum amount of sample (volume) is a very important parameter of chromatographic separations. The larger the volume of sample to be chromatographed, the bigger the amount of substance that can be taken for separation. On the other hand, several authors have observed the decrease of separation characteristics with increased sample volume (11-13). The solution of the problem of how to avoid the decrease of efficiency of chromatographic column, depends on the applied technique of liquid chromatography.

Kirkland (11) suggested limitation of sample size from the point of view of observed efficiency, that is to limit the sample size to such an amount over which the efficiency shows a significant decrease.

On the other hand, the number of theoretical plates (N)vs log sample size diagram shows a conspicious inflection and the end of upper constant line can be easily determined by the intersection tangents of the inflection and upper constant line. The value of sample size belonging to the inflection means a lower limit, the end of upper constant line (determined by the intersection) gives an upper limit (Fig.1).

While these studies suggest keeping the efficiency above a certain value by limitation of sample size, another point of view proposes the improvement of resolution by recycling. Both gel permeation with organic solvent eluent (14-16) and gel chromatography in aqueous solution (17-18) allow the resolution to be increased by recycling even in the case when the low value of resolution is caused by large sample size (14).

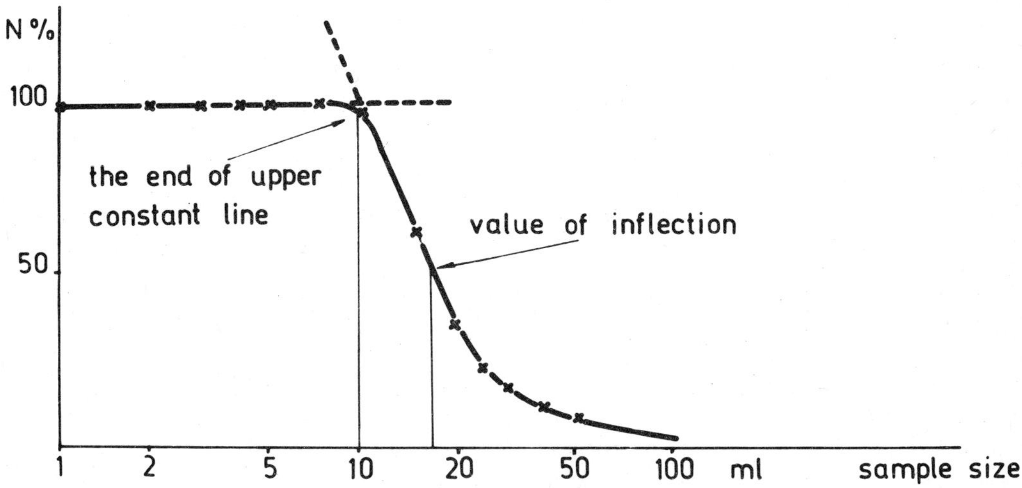

Figure 1. Limitation of the sample volume on the basis of theoretical plate number (N) vs log sample size diagram. Either the end of the upper constant line or the value of inflection can give the limit of the sample size.

Similar point of view is well explained in the papers (19-21) which consider the resolution between peaks to be separated, not the efficiency. The separation usually means a certain result in that the peaks need to be far enough from each other, e.g., the resolution of several given peaks has to reach an adequate value. From this consideration, group separation allows much larger sample size than fractionation. This topic was not only suggested but also detailed by Fischer (19), Kremmer and Boross (20) derived from Pharmacia brochure (21) which gives the sample limit according to the eqn.:

$$V_{sample} \lll V_{sep} \cong V_{e_B} - V_{e_A}$$

where V_{e_B}, V_{e_A} are the elution volume of B and A components, V_{sample} is the sample size and V_{sep} is the volume between the separated peaks, respectively. The same source states that in desalting and buffer exchange, volumes up to approximately 30% of the total bed volume can be used to minimize dilution and still retain good separation. Later, it (20) gives the sample volume as 1-5 % of the bed volume in the case of fractionation and it points out also that the smaller volume does not normally improve resolution. At the same time, the division of gel chromatography into "group separation" and "fractionation" subgroups reflects only the most schematic classification. Investigation of particular problems suggests that the individual limit of sample size on the previously obtained chromatogram of gel chromatographic separation can present the method of choice.

The majority of gel chromatography allows the isolation of substances which have been separated and monitored. The isolation means first of all the selection of a certain part of the effluent to be collected together. This process is usually performed by fraction collectors which work automatically. However, the basic question, which size of effluent volume is to be collected remains unanswered. In the literature (22-24) the problem of fractionation or collecting fractions has been mainly restricted to the operative aspects of fraction collectors themselves. Only some recently published works (25-26) mentioned that the individual peaks as well as transitional zones should appear in 3-5 fractions at least in the case of fractionation or altogether several fractions are collected in the case of group separations.

A very detailed discussion of the separation of chromatographic peaks and its demonstration was published by Saunders (27) who supplied the chromatographic profiles of wide variation of partially or totally resolved peaks. The extent of contamination and the degree of purity were also given in his paper (27).

Svenson (28) has also dealt with the subject of the resolving zones. The analog resolving was taken from the topic of light scattering and Svenson presented the possible definition of separation in the cases of partially or totally resolved peaks having the same or different peak height.

On the other hand, the joint discussion of separation and fractionation or, more precisely, the <u>sample size-separation-</u>

fractionation problem seems to be lacking from the literature. Nevertheless this topic is of great importance in any kind of preparative separations. The number of fractions is one of the most important characteristics of the gel chromatographic separations of preparative nature. The chromatographic process yields the continuous change of the effluent, but only a few characteristics of the effluent can be monitored directly (UV extinctions, fluorescence, conductivity,etc.). Numerous and basic parts of characteristics of the substances (biological activity, enzyme activity,etc.) can be detected only from the collected fractions. The process of fractionation never improves the separation but decreases very often. At the same time, the larger the number of fractions, the smaller the remixing of the separated compounds. This phenomenon seems to influence us in choosing as high as possible number of fractions. A high number of fractions, however, causes the increase of analytical work in the investigation of the individual fractions, the greater risk in loss of substance during the change of fractions, and at analysis, i.e.the high number of fractions comprehends both the increased work of indirect monitoring and diminishing the recovery. From this point of view, the number of fractions is advised to be limited as low as possible.

Before the detailed discussion of the fractionation some definitions and conditions should be stated.

Peak-cutting fractionation means the case when one fraction collects the effluent until the peak maximum and the other fraction does it from the maximum.

Peak-keeping fractionation occurs when the peak maximum coincides with the half of the fraction. These two basic cases are given in Fig.2.

Basic condition of the calculation is that the elution peaks are symmetrical and can be approximated by the Gaussian formula. The discussions underline the isolation of a main (single) component and its separation from the neighborhooding peaks (from the previous and following but nearest eluting substances) are to be considered. In this case, the basis of any calculation is the peak to be isolated. In any other cases,the component eluting first is to be taken into consideration. The given extinctions are essentially proportional to concentrations, that is the dilution during fractionation does not cause the basic change of specific extinction coefficient. The calculation of the numerical average of extinctions performs in a practical way the calculation of the numerical average of the substance content.

However, the simulation of fractionation does not substitute for the checking of fractions for their purity.

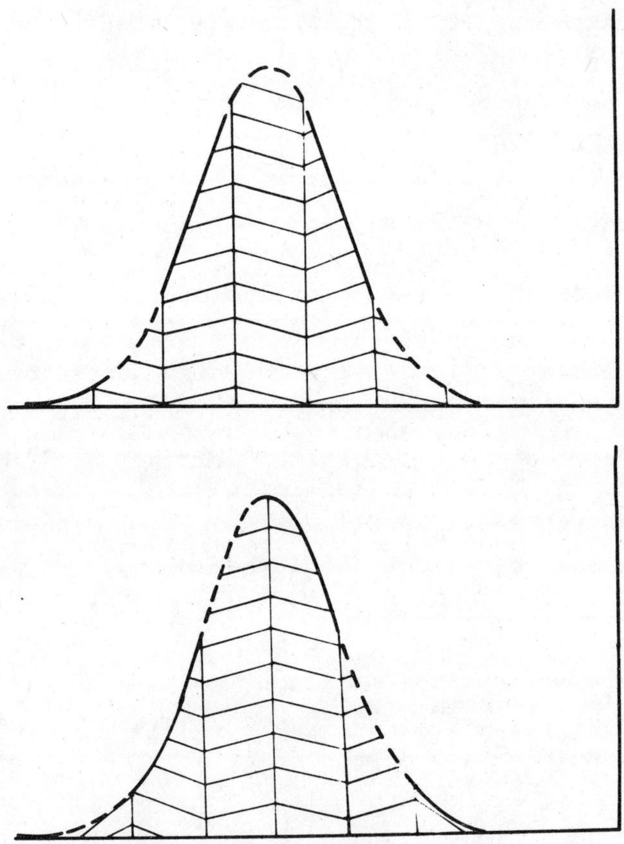

Figure 2. Two different kinds of fractionation, that is the peak-keeping (upper curve) and the peak-cutting (bottom curve) modes of fractionation.

In our earlier work (29) the collecting of sigma fractions was suggested, that is sectioning the elution profile according to sigma. This is demonstrated in Fig.3.

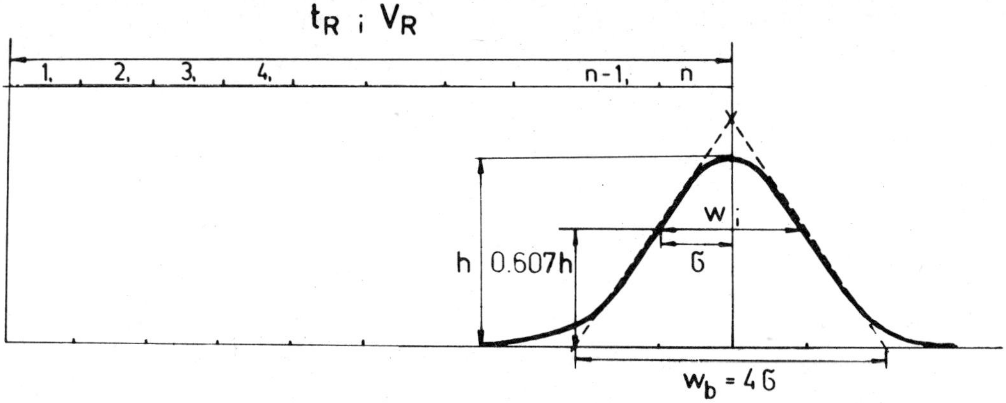

Figure 3. Sectioning the elution curve according to the standard deviation (sigma) of the Gaussian type of peak.

It means the easy definition of the extent of fractionation on the one hand, but on the other hand the calculation of the fraction number to be collected is also simple.

If the period before the elution peak center is divided into sigma parts, the occurrence of such segments is the fraction number (n_{fr}):

$$n_{fr} = \frac{t_R}{\sigma t} = \frac{V_R}{\sigma V} .$$

This value is identical with the square root of theoretical plate number (N) of the chromatographic column according to its definition:

$$\sqrt{N} = \frac{t_R}{\sigma t} = \frac{V_R}{\sigma V} \quad ; \sqrt{N} = \frac{t_R}{\sigma t} = \frac{V_R}{\sigma V} = n_{fr}$$

Performing this fractionation, the size of each fraction is equal to the standard deviation in time (or volumetric) units, that is equal to one fourth of the peak width at the baseline concerning the peak to be separated and isolated.

In our earlier works (30-33) we discussed the approximation of gel chromatographic elution curves that were obtained from aqueous gel chromatography on Bio-Gel P-2 and investigated some indolic and phenolic compounds. Our fundamental findings were:

1. The elution curves are symmetrical and can be approximated by the well known Gaussian formula:

$$Y_i = A \exp \left[- \frac{(X_i - X_o)^2}{2\sigma^2} \right]$$

2. The curve calculated in this way and the experimental curve show a good agreement. Although the substances eluting after strong retardation (aromatic adsorbtion, hydrophobic interactions, etc.) show nice symmetry, they can be surprisingly well approximated.

3. The data are presented in units of reverse transmission (100-T%). Although the scale of photometer used (UVICORD II) is given in transmission, the maximal value of each chromatographic curve is at the minimum value of transmission.

4. The resulting points of several elution (calculated) curves can not be obtained by direct addition, multiplication or substraction as neither the value of transmission nor reverse transmission is proportional with the concentration. Each value of the elementary reverse transmissions has to be transformed into extinction (which is proportional with concentration) summed and the summed extinction values need to be transformed into reverse transmission in order to generate the experimental curve. The addition of five curves successfully reproduced the experimental data if the curves had the same peak width and peak height but they differed from each other in the position of peak maxima. On the other hand, the addition of three curves sufficed if the above mentioned charachteristics of the chromatographic curves were totally different.

5. Five superimposed peaks can be used for simulating the increase of sample size (31-33) and summation of three peaks can be used for calculation of three, partially overlapping chromatographic curves. The expansion of the number of curves to be added together was limited both by the number of calculation steps and the number of registers. To develop further our calculation potential, we modified our basic program and used the secondary registers of HP-97 as well.

In our paper we demonstrate the numerical calculation of the elution profile from numerous elementary peaks, we outline the calculation of sample size limit experimentally as well as numerically. Furthermore, we suggest the method of numerical approximation of fractionation and we supply easy programs for these above mentioned calculations with a Hewlett Packard HP-97 calculator. In connection with the laboratory work, we expand the calculations of gel chromatographic separations by setting a joint program for simulation of both sample size effect and fractionation together.

MATERIALS AND METHODS

Materials:

All substances were commercially available. Bio-Gel P-2, 100-200 mesh (wet), control No. 187362, exclusion limit 1,800 daltons was purchased from Bio-Rad Laboratories (Richmond, Calif., USA), phenylethylamine purum, tyramine CHR, dopamine purum, triptamine purum, all hydrochlorides, benzoic acid purum, 4-hydroxy benzoic acid purum and 3,4-dihydroxy benzoic acid CHR were purchased from Fluka AG (Buchs, Switzerland):serotonine creatine sulfate, chromatographically pure, was obtained from Mann Research Laboratories, Inc. (New York, USA); bovine serum albumin was obtained from Phylaxia (Budapest, Hungary).Sephadex G-15 (trade mark name of the Pharmacia produced gel filtration material), lot. No. 7755, particle size 40-120 μm, was obtained from Pharmacia Fine Chemicals (Uppsala, Sweden).

Methods:

Bio-Gel P-2 Sephadex G-15 gels were supplied as in the form of dry powder. The gel was added to excess of water with constant strirring and then allowed to swell for 24 hours. The mixture of swollen gel and water (1:1) was poured into the column and the gel was allowed to settle for 15 minutes. The outlet was then opened and the further filling was performed during constant flow. The gel-water suspension was continuously supplied. After the settled gel had filled 90-95 percent of the column, the upper flow adaptor was attached and the elution was started and continued for 48 hours with 0.9 % sodium chloride solution (physiological saline solution). The samples were applied through the flow adaptor, peristaltic pump and capillary tubes by the transfer of the inlet tube from the eluent reservoir to the sample container. The start of elution was indexed by the change of the baseline of Uvicord II recorder.

Apparatus:

Various parts of the ReCyChrom System (LKB-Produkter, Bromma, Sweden) were used. These included the Peristaltic Pump, Uvicord II with Recorder and Combi Cold Rac. Chromatographic Columns were purchased from Pharmacia Fine Chemicals (Uppsala, Sweden). Columns in the Combi Cold Rac were kept at + 4°C.

The calculations were made on a Hewlett-Packard HP-97 calculator.

RESULTS

Approximation of gel chromatographic curves

The experimental elution profile for the separation of substituted benzoic acids is given in Fig.4/a. The peaks are symmetrical and this fact suggests that the peaks may be approximated by the Gaussian formula. To facilitate calculations we developed a program for the Hewlett-Packard HP-97 calculator. The scheme for the process of calculation from point to point of the elution curve can be seen in Table 1. This program (detailed in Table 2) allows the calculation of the points of three individual curves (No. 1 curve with LBL 1, steps 027-045; No. 2 curve with LBL 2, steps 046-064; No. 3 curve with LBL 3, steps 065-083), the transformation of the reverse transmission values into extinction (LBL a in steps 084-095) including the summation of the calculated and transformed values (ST + 3 in step 094) as well as the retransformation of the sum into reverse transmission (LBL b in steps 096-116). The program serves for the calculation of the whole point series since the end of calculation of any point is followed by the change of the place of calculation that is stepped further (steps 004-010) if the calculated value is not under 0.001 (steps 018-026).

Table 3 consists of the arrangement of characteristics of three curves and parameters of calculation in the registers.

The program given in Table 2 has been used for calculation of elution curves given in Fig.4/b. The comparison of the experimental curve by the calculated point series is demonstrated in Fig.5 where the experimental curve is sandwiched between two calculated ones which have 10 % larger peak maxima and standard deviation as well as 10 % smaller peak maxima and standard deviation, respectively, than the experimental elution curve.

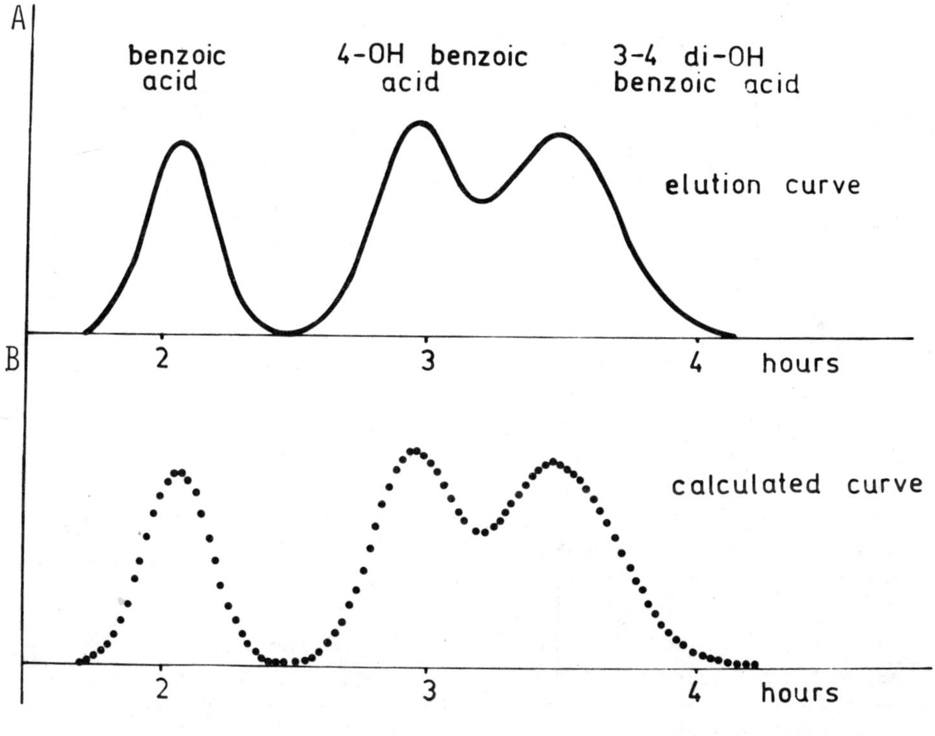

Figure 4. Elution profile of separation of substituted benzoic acids (top curve) and the calculated values (bottom point series) by the use of the calculation program given in Table 2. Gel chromatography was performed on Bio-Gel P-2 (90 cmx2.5cm) gel column using 0.9 % sodium chloride solution as eluent.

TABLE 1.
The process of calculation of the elution curve

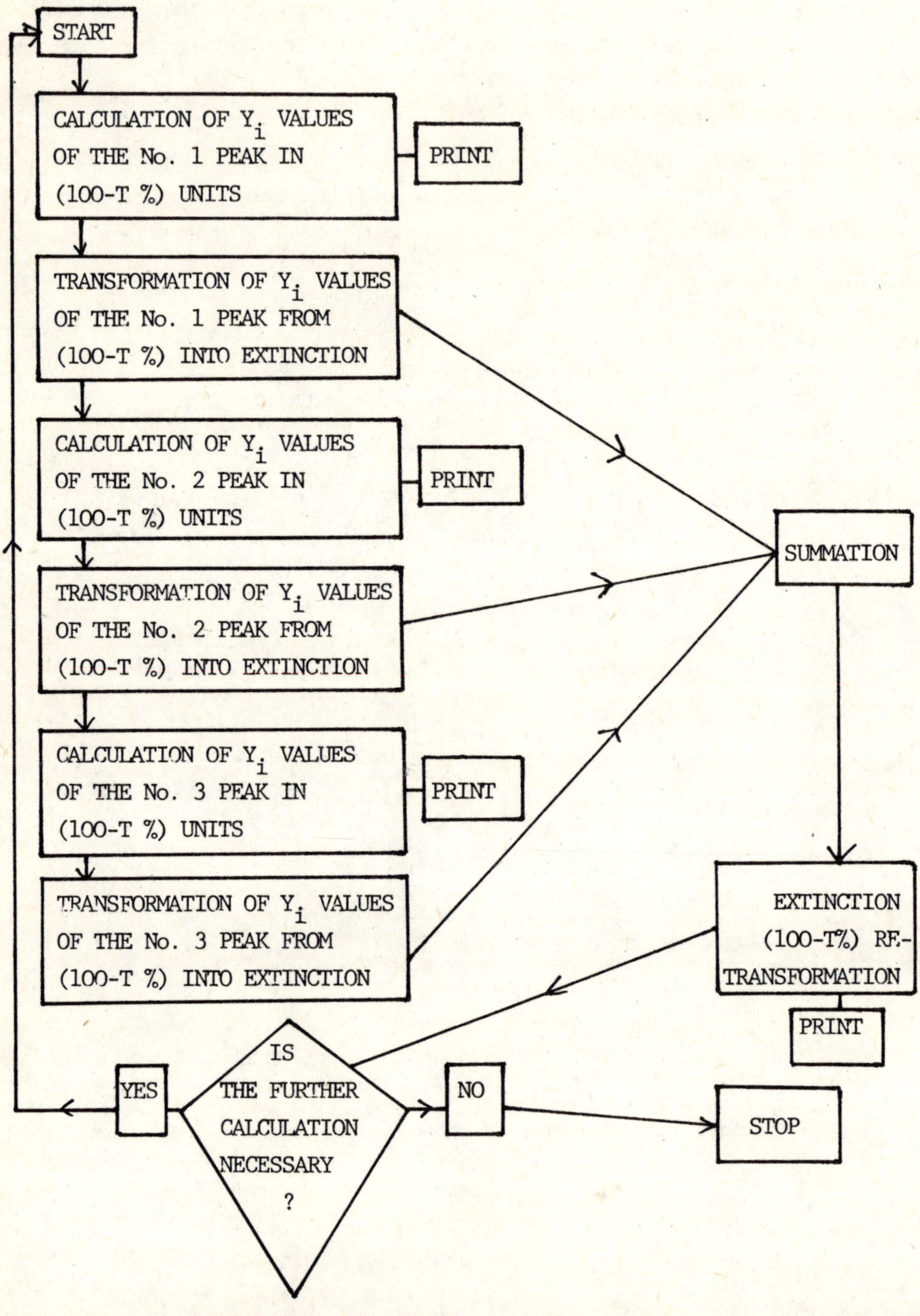

TABLE 2.

Individual steps of calculation program

001	*LBLA	040	x	079	1
002	CLX	041	1	080	.
003	STO3	042	.	081	2
004	RCLØ	043	2	082	:
005	ENT↑	044	:	083	RTN
006	RCL2	045	RTN	084	*LBLc
007	+	046	*LBL2	085	1
008	STO2	047	RCL2	086	0
009	SPC	048	RCL5	087	0
010	PRTX	049	-	088	X⇄Y
011	GSB1	050	X^2	089	-
012	GSBa	051	RCL8	090	LOG
013	GSB2	052	X^2	091	2
014	GSBa	053	2	092	X⇄Y
015	GSB3	054	x	093	-
016	GSBa	055	:	094	STO+3
017	GSBb	056	CHS	095	RTN
018	Ø	057	e^X	096	*LBLb
019	.	058	RCLB	097	RCL3
020	Ø	059	x	098	10^X
021	Ø	060	1	099	1
022	1	061	.	100	0
023	X⇄Y	062	2	101	0
024	X > Y?	063	:	102	X⇄Y
025	GSBA	064	RTN	103	:
026	R/S	065	*LBL3	104	1
027	*LBL1	066	RCL2	105	0
028	RCL2	067	RCL6	106	0
029	RCL4	068	-	107	X⇄Y
030	-	069	X^2	108	-
031	X^2	070	RCL9	109	1
032	RCL7	071	X^2	110	.
033	X^2	072	2	111	2
034	2	073	x	112	x
035	x	074	:	113	DSP8
036	:	075	CHS	114	PRTX
037	CHS	076	e^X	115	DSP2
038	e^X	077	RCLC	116	RTN
039	RCLA	078	x	117	R/S

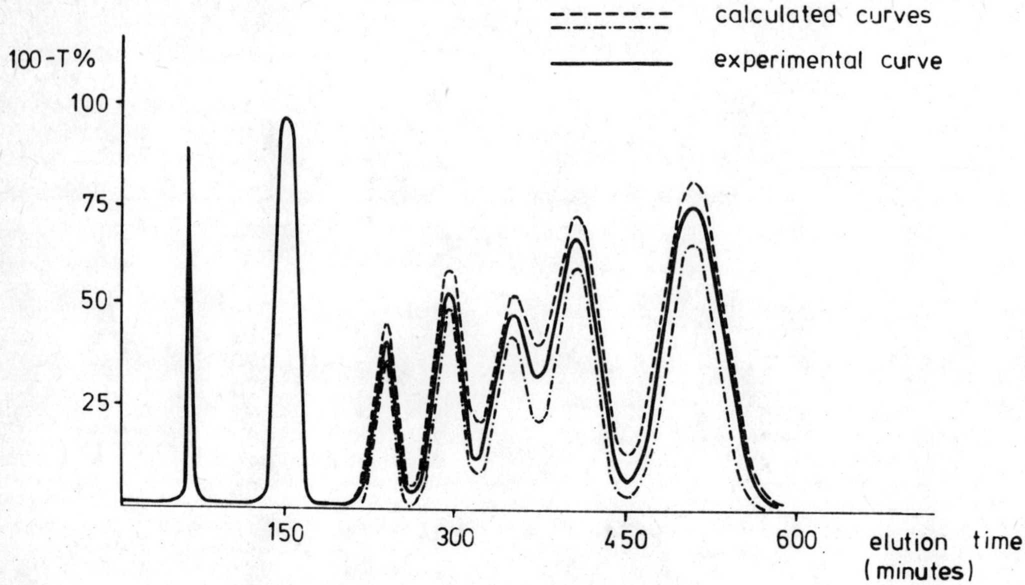

Figure 5. Demonstration of the fittness of the calculated curves to the experimental one. The experimental chromatogram (solid line) shows the separation of 60 mg albumin, 0.3 ml acetone, 35 mg phenylethylamine, 35 mg tyramine, 35 mg dopamine, 10 mg serotonine and 10 mg tryptamine on Bio-Gel P-2 column (90 cm x 2.5 cm) using 0.9 % sodium chloride solution as eluent.

Experimental elution curve is sandwiched between the two calculated ones which have 10 % larger peak maxima and standard deviations (dotted line) as well as 10 % smaller peak maxima and standard deviations (dashed line), respectively, than the experimental curve.

Simulation of enlarging the sample size by calculation

The approximation of the effect of the increase on the sample size was initiated in our earlier works (31-32). The modelling of the effect has succeeded; the components or elution peaks can be followed one by one. It was caused by the fact

Table 3. Arrangement of characteristics of three elution curves and parameters of calculation in the registers.

REGISTERS	PARAMETERS
0	Length of steps
1	Blank
2	Start (and place) of calculations
3	Blank
4	Center of No. 1 peak
5	Center of No. 2 peak
6	Center of No. 3 peak
7	Standard deviation of the No. 1 peak (σ_1)
8	Standard deviation of the No. 2 peak (σ_2)
9	Standard deviation of the No. 3 peak (σ_3)
A	Height of the No. 1 peak
B	Height of the No. 2 peak
C	Height of the No. 3 peak

that the deconvolution of a single elution peak into five elementary curves and their superposition exhausted the possible steps of the program and the possible places of primary register up to their two-third. In some cases the shape of the approached curves was appropriate but in other cases the greater number of elementary curves would have given fairly good results. The phenomenon is demonstrated in Figs.6 and 7.

In addition to the better shape of calculated curves, the further problem to be solved is the construction of a calculated elution profile which consists of several chromatographic peaks, each resulting from 10-12 elementary curves. This development needed either 10-20 times more calculation steps and 5 times larger capacity of registers or the basic alteration of the calculated system. This latter possibility means the use of fewer subroutines by using the same subroutine for calculation of different curves which has been realized by

```
*LBLA
CLX
STO1
RCLD
RCLØ
+
STOØ
PRTX
RCL4
STOA
RCL5
STOB
RCL6
STOC
GSB1
GSB2
GSB2
GSB2
GSB2
GSB2
GSB2
GSB2
GSB2
GSB2
GSB2
GSB2
GSB2
RCL7
STOA
RCL8
STOB
RCL9
STOC
GSB1
GSB2
GSB2
GSB2
GSB2
GSB2
GSB2
GSB2
GSB2
GSB2
GSB2
GSB2
P⇄S
RCL1
STOA
RCL2
STOB
RCL3
STOC
P⇄S
GSB1
GSB1
GSB2
GSB2
GSB2
GSB2
GSB2
GSB2
GSB2
GSB2
GSB2
GSB2
GSB2
GSB2
P⇄S
RCL4
STOA
RCL5
STOB
RCL6
STOC
P⇄S
GSB1
GSB2
GSB2
GSB2
GSB2
GSB2
GSB2
GSB2
GSB2
GSB2
GSB2
GSB2
GSB2
P⇄S
RCL7
STOA
RCL8
STOB
RCL9
STOC
P⇄S
GSB1
GSB2
GSB2
GSB2
GSB2
GSB2
GSB2
GSB2
GSB2
GSB2
GSB2
GSB2
GSB2
GSBa
GSBc
RTN
*LBL1
RCLE
1
2
x
RCLA
+
STOA
RTN
*LBL2
RCLA
RCLE
-
STOA
RCLØ
-
X^2
RCLB
X^2
2
x
:
CHS
e^X
RCLC
x
1
.
2
:
1
Ø
Ø
X  Y
-
LOG
2
X⇄Y
-
ST+1
ST+2
RTN
*LBLa
RCL1
10^X
1
Ø
Ø
X⇄Y
:
1
Ø
Ø
X⇄Y
-
1
.
2
x
DSP4
PRTX
DSP2
CLX
STO1
RTN
*LBLc
1
ST+3
RCLI
RCL3
-
X=Y?
GSBd
GSBA
RTN
*LBLd
RCLØ
DSP7
PRTX
RCL2
RCL3
:
10^X
1
Ø
Ø
X⇄Y
:
1
Ø
Ø
X⇄Y
-
1
.
2
x
DSP9
PRTX
DSP2
P⇄S
RCLØ
P⇄S
RCLØ
-
X<Y?
R/S
CLX
STO2
STO3
GSBA
RTN
```

←

TABLE 4.

Calculation program for approximation of 5 x 12 elementary curves, their superposition and the fractionation

the change of peak parameters with the help of the main program. The schematic program of calculation of k x n (e.g. 5 x 12) curves and their superposition is shown in Fig.8. Just before the calculation of the first curve, the center point of the peak is dislocated with n-times of the shift e.g. the extreme right elementary curve was calculated first. Later the second, third, etc. curve (from right) were calculated. Finally the curve having the original peak center was calculated.

Similarly, each individual chromatographic peak was calculated by constant shifting. As the sample size is the same for any component of the sample to be or having been separated, the constant shift is adequate for any component. The summarized value of shifts composes the volumetric size of the sample as it was demonstrated earlier (33).

The program elaborated for Hewlett-Packard HP-97 calculator is given in Table 4 while the arrangement of data and characteristics of the calculation in the registers are given in Table 5. In this program, the substitution of characteristics of No. 1.,2.,3.,4. and 5. curves is carried out in steps 009-014, 028-033, 047-054, 068-075 and 089-096, respectively.

By using the program given in Table 4, we calculated the effect of sample size. The chromatographic peaks were separated into 12 elementary curves each (addition of which gives back the original peaks). The constant shift was placed into register E. Original (experimental) and calculated chromatograms can be seen in Fig.9.

Fig. 10 shows that the sample volume can be increased to 50 ml in the case of 90 cm x 2.5 cm (442 ml total volume) Bio-Gel P-2 column and still separated the mentioned sample.

TABLE 5.

Arrangement of the parameters in the P and S registers of the Hewlett-Packard HP-97 calculator

Parameters in P registers	Parameters in S registers
0 Start of calculation	0 End of calculation
1 Blank	1 Center of the 3rd curve
2 Blank	2 Standard deviation of the 3rd curve
3 Blank	3 Peak height of the 3rd curve
4 Center of the 1st curve	4 Center of the 4th curve
5 Standard deviation of 1st curve	5 Standard deviation of the 4th curve
6 Peak height of the 1st curve	6 Peak height of the 4th curve
7 Center of the 2nd curve	7 Center of the 5th curve
8 Standard deviation of the 2nd curve	8 Standard deviation of the 5th curve
9 Peak height of the 2nd curve	9 Peak height of the 5th curve

A Blank

B Blank

C Blank

D Length of steps

E Shift of furves

I Number of steps for each fraction

Calculation of fractionation

Using the above detailed logic line of calculations (see Table 4 and Fig.8), the scheme can be extended to the simulation of fractionation, too. The procedure will consists of the following steps:

1. calculation of the individual members of the point series in (100-T %) units,
2. transformation of the calculated values into extinctions,
3. performing the average of extinctions,
4. transforming back the averaged extinctions into (100-T %) units

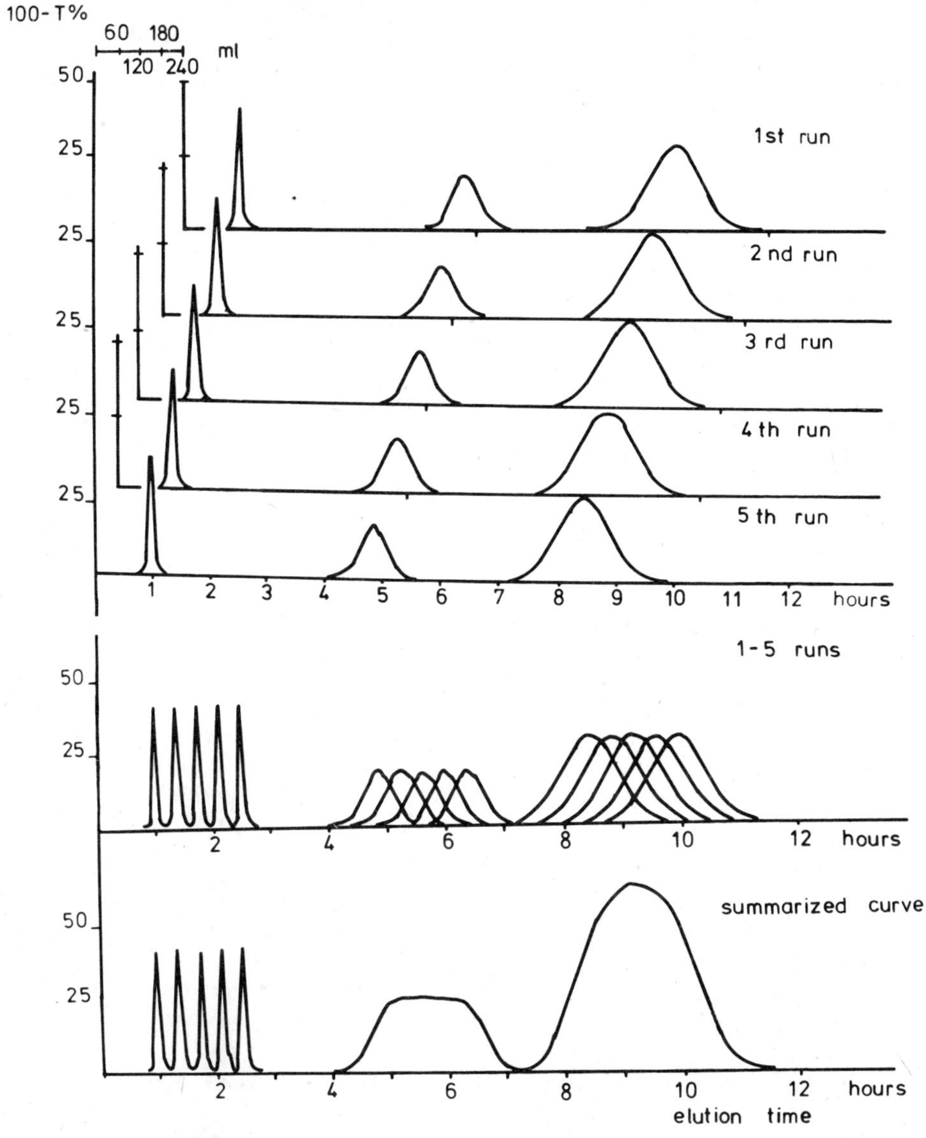

Figure 6. Modelling of sample size by the construction from 5 elementary curves in the cases of albumin (1st eluting peak), phenylalanine (2nd peak) and tryptamine (3rd peak). The curves of tryptamine and phenylalanine have become well simulated but the albumin peak needs much higher number of elementary curves.

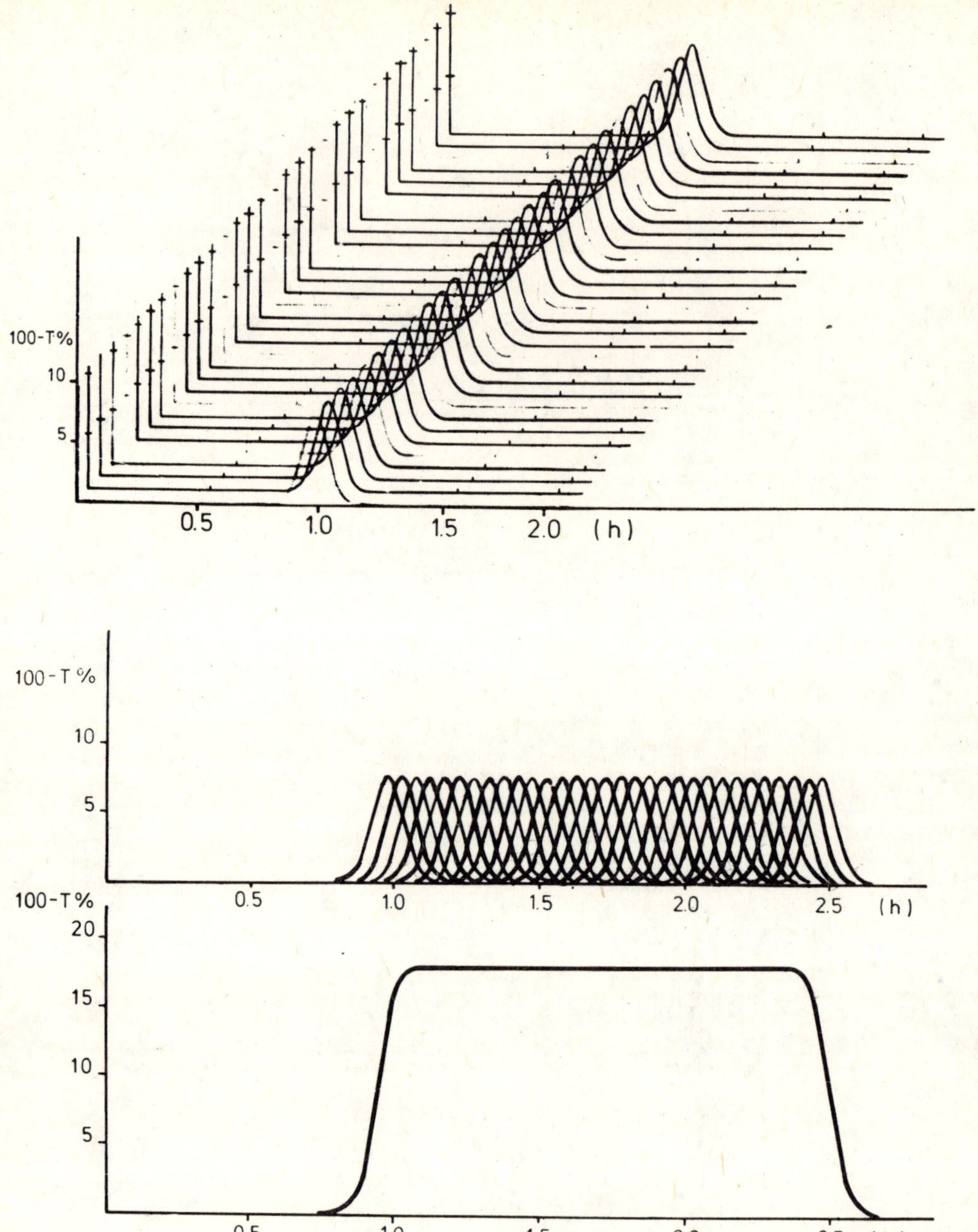

Figure 7. Albumin peak and its construction from 31 elementary curves.

Calculation of the points of the No.1. peak at x_i

dislocation of the peak center with (n) times of the shift from left to right,

calculation of the point of the (n)th elementary curve
calculation of the point of the (n-1)th elementary curve
calculation of the point of the (n-2)th elementary curve

calculation of the point of the 2nd elementary curve
calculation of the point of the 1st elementary curve

Calculation of the points of the No.2. peak at x_i

dislocation of the peak center with (n) times of the shift from left to right,

calculation of the point of the (n)th elementary curve
calculation of the point of the (n-1)th elementary curve
calculation of the point of the (n-2)th elementary curve

calculation of the point of the 2nd elementary curve
calculation of the point of the 1st elementary curve

Calculation of the points of the No.k. peak at x_i

dislocation of the peak center with (n) times of the shift from left to right,

calculation of the point of the (n)th elementary curve
calculation of the point of the (n-1)th elementary curve
calculation of the point of the (n-2)th elementary curve

calculation of the point of the 2nd elementary curve
calculation of the point of the 1st elementary curve

Calculation of the points of the No.1. peak at x_{i+1}

Figure 8. The train of calculation of a simulated chromatogram consisting of n x k elementary curves.

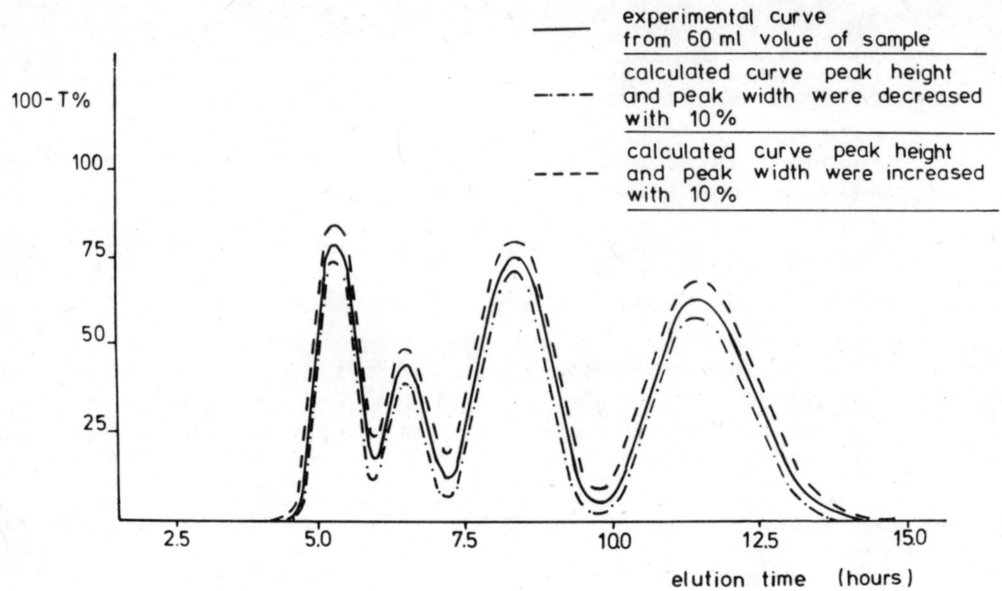

Figure 9. The calculated elution curves which are sandwiching the experimental one for the separation of satietin on a 2.5 x 90 cm Bio-Gel P-2 column using 0.9 % sodium chloride solution as eluent and monitored the separation at 254 nm.

The process can be programmed into the Hewlett-Packard HP-97 calculator. The fractionation is performed by the summation of the elementary values (ST+2) up to the end of a certain period (steps 178-187 in subroutine LBLc). The operation is completed by the division of the sum with the number of elementary points and retransformation of this (average extinction) into (100-T%) as well as the end of the program if the last average is under a certain value given in the secondary register No.1.

Using the above detailed calculation process, the fractionation can be performed according to sigma half, sigma and two sigma if simulated chromatographic peaks with different resolutions to each other are presented.

Fig. 11 shows sigma half, sigma and two sigma fractionation of such peaks which show 3,4,5 and 6 sigma resolution with the peak-keeping fractionation. In Fig.12 the peak-cutting fractionation is presented.

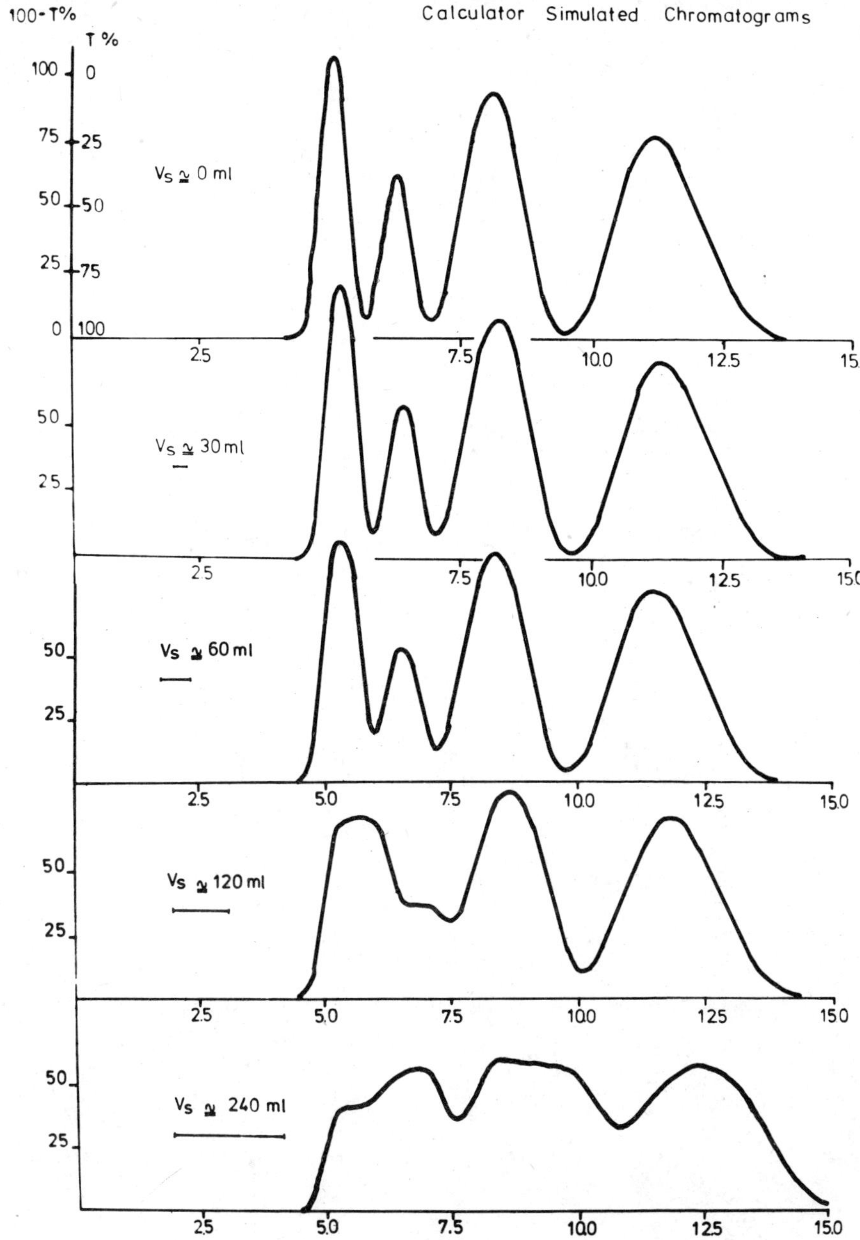

Figure 10. Calculated simulated chromatograms indicate the sample size effect for the separation of satietin. The same amount of the substance to be separated was in each sample having very small, 30,60,120 and 240 ml volume.

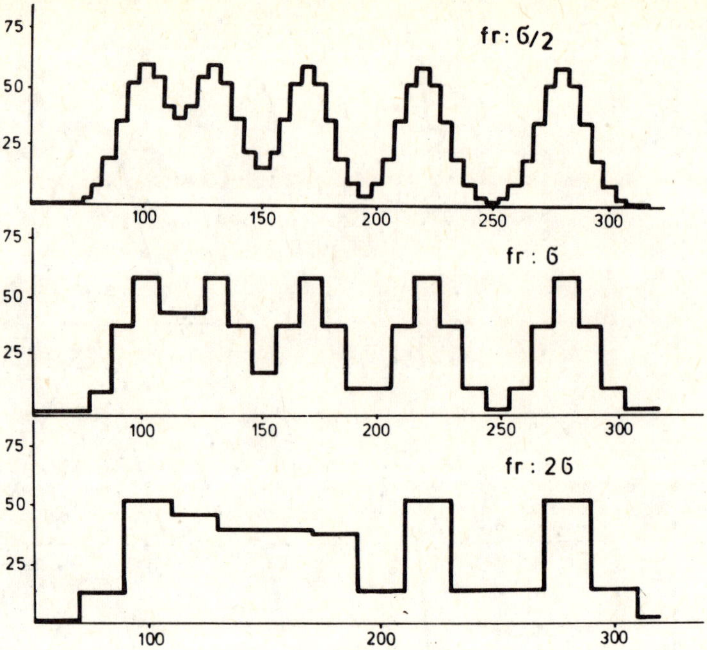

Figure 11. Sigma half, sigma and two sigma fractionation of a peak-series having 3,4,5 and 6 sigma resolutions. It is the peak-keeping mode of fractionation for the 1st peak.

The comparison of Figs. 11 and 12 shows that the size of fractions plays a major role and not its arrangement (that is peak-keeping or peak-cutting fractionation).

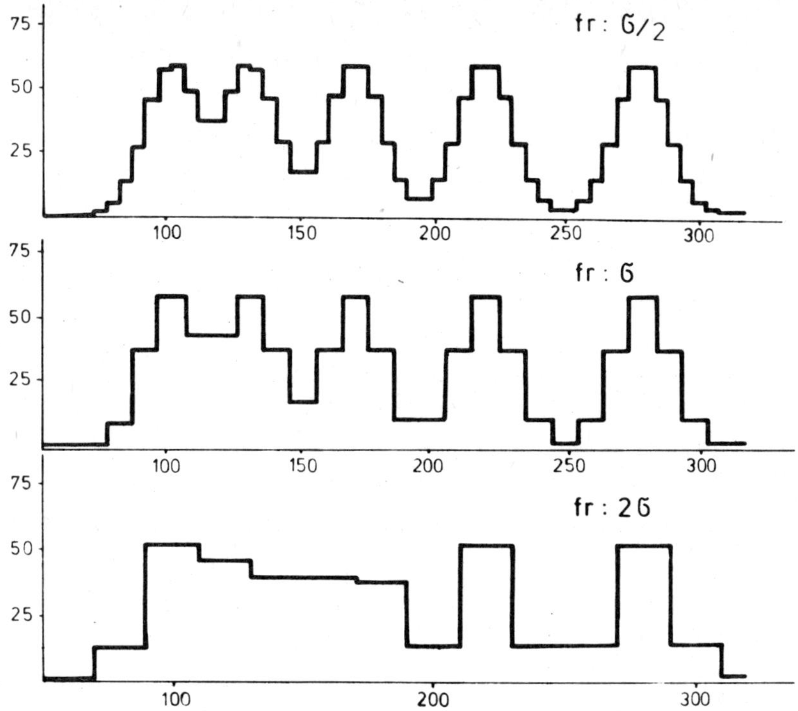

Figure 12. Sigma half, sigma and two sigma fractionation of a peak-series having 3,4,5 and 6 sigma resolutions. It is the peak-cutting mode of fractionation for the 1st peak.

DISCUSSION AND CONSIDERATIONS

The gel chromatographic separation of several phenyl- and indole-alkyl compounds gave Gaussian type of peaks. These curves can be numerically approximated by the use of the Gaussian formula and the process can be performed with a Hewlett-Packard HP-97 calculator. Both the single elution curve and the elution profile of separation of several components can be approximated with good fitting. The increase of sample size could also be constructed and the calculated and experimental curves show fairly good agreement.

Some cases show that the sample size limitation by the help of calculation gives the adequate results; that is, the calculation is able to indicate the sample size which does not affect the resolution of individual peaks. Moreover, the mathematical calculations of sample size limitation reduces the use of the apparatus except for the first preliminary experiment. The saving of separable compounds, time, etc. is also considerable.

The effect of increased sample size on the real chromatographic profile is also more pronounced and expressive than the consideration of the change of some chromatographic characteristics as resolution, number of theoretical plates,etc.

In addition to the practical results, the progress of calculation is simple. The figures demonstrate how surprisingly large volumes can be sampled without observable devaluation of separation, i.e., the change of chromatographic profile occurred at a moderetely large volume of sample.

At the same time, the isolation of different components can be facilitated if the desired results have been reached by the separation and the effect of fractionation does not disturb the resolution, i.e., the remixing of the previously separated and resolved peaks is avoided.

The basic suggestion has been the so called sigma fractionation, that is to collect the eluent into sigma segments as the standard deviation of the peak to be isolated.

The formula of fraction number, i.e. the number of fractions to be collected, is equal to the square root of the theoretical plate number. It gives an easy way of defining the fraction number. This type of fractionation seems to be adequate, however, for most separations both the fraction number and the remixing effect of the totally or partially separated components have been minimized.

Two different types of the separations served to demonstrate the effect of sigma fractionation:

1. The simulated fractionation of 3,4,5 and 6 sigma resolution curves,
2. The separation of a substance of natural origin (34) called satietin.

All figures demonstrate the correctness of the statement that a general suggestion may be the sigma fractionation which practically keeps the results of separations but limits the number of fractions at a moderately low value.

REFERENCES

(1) BUYS,T.S.and de CLERK,K., J.Chromatogr.67 1 /1972/
(2) BUYS,T.S. and de CLERK,K., J.Chromatogr.67 13 /1972/
(3) BUYS,T.S. and de CLERK,K., J.Chromatogr.69 87 /1972/
(4) de CLERK,K., BUYS,T.S. and PRETORIUS,V., Separation Science 6 733 /1971/
(5) de CLERK,K. and BUYS,T.S., J.Chromatogr.63 193 /1971/
(6) JOHNSON,J.F. and PORTER,R.S. (Eds.): Analytical Gel Permeation Chromatography. Interscience,New York,1968.
(7) PIERCE,P.E. and ARMONAS,J.F., J.Polymer Sci.21 23 /1968/
(8) TUNG,L.H., J.Appl.Polymer Sci., 10 375 /1966/
(9) MEYERHOFF,G., J.Polymer Sci. 21 311 /1968/
(10) PICKETT,H.F., J.Polymer Sci.21 67 /1968/
(11) KIRKLAND,J.J.: The Practice of Liquid-Liquid Chromatography.in: Kirkland,J.J.(Ed.) Modern Practice in Liquid Chromatography. p.175. Wiley Interscience,New York, 1971.
(12) VERZELE,M. and GEERAERT,E., J.Chromatogr.Sci. 18 559/1980/
(13) COQ,B., CRETIER,G. and ROCCA,J.L., J.Chromatogr.,186 457 /1979/
(14) BOMBAUGH,K.J., J.Chromatogr.53 27 /1970/
(15) BOMBAUGH,K.J. and LEVANGIE,R.F., J.Chromatogr.Sci. 8,560 /1970/
(16) MARTIN,M., EON,C.,VERILLON,F. and GUIOCHON,G., J.Chromatogr. 125,17 /1976/
(17) KALÁSZ,H., J.Chromatogr. 78 233 /1973/
(18) KALÁSZ,H. NAGY,J. and KNOLL,J., J.Chromatogr. 107 35 /1975/
(19) FISCHER,L.: An Introduction to Gel Chromatography,North Holland, Amsterdam 1971.
(20) KREMMER,T. and BOROSS,L.: Gel Chromatography. Wiley Interscience, New York and Akadémiai Kiadó,Budapest, 1979
(21) ANONYMUS: SEPHADEX,R. Gel Filtration in Theory and Practics. (Pharmacia Fine Chemicals, Uppsala,Sweden,1966)
(22) DETERMANN,H.: Gelchromatographie, Springer, Berlin,1967.
(23) MIKES,O. (Ed.): Laboratory Handbook of Chromatographic and Allied Methods. Ellis Horwood, Chichester,G.B.,1979
(24) HEFTMANN,E. (Ed.): Chromatography,Reinhold, New York,1967
(25) KREMMER,T. and BOROSS,L.: Gel Chromatography (in Hungarian) Muszaki Konyvkiadó, Budapest, 1974.
(26) NICE,E.C., CAPP,M. and O'HARE,M.J., J.Chromatogr. 185 413 /1979/
(27) SAUNDERS,D.L.: Techniques in Liquid Column Chromatography. in: Chromatography (Ed.: Heftmann,E.).Van Nostrand-Reinhold, New York, 1975
(28) SVENSSON,H., J.Chromatogr. 25 266 /1966/

(29) KALÁSZ,H., Proc. 17th Hungarian Ann.Meet.Biochem.(Kecskemét Hungary, 1977/ p. 161.
(30) KALÁSZ,H.,NAGY,J. and KERECSEN,L.: Gel Chromatography of Phenylalkylamines. in: Monoamine Oxidase and their Selective Inhibition (Eds.: Knoll,J. and Magyar,K.). Pergamon Press, New York and Akadémiai Kiadó,Budapest, 1980.
(31) KALÁSZ,H.,KERECSEN,L., KNOLL,J. and MAGYAR,K., Proc. Internat Symp.Biomed.Applic. Chromatography (Castle Hluboka, Czechoslovakia, 1978).
(32) KALÁSZ,H.: Gel Chromatography. in:Methods in Protein Analysis (Ed.:Kerese,I.) Akadémiai Kiadó, Budapest and Ellis Horwood, Chichester,G.B., in press.
(33) KALÁSZ,H.: Optimal Possibilities of Gel Chromatography and its Application for Separation of Biologically Active Endogenous Compounds. Dissertation.Budapest,1979.
(34) KNOLL,J., KALÁSZ,H.,KNOLL,B. and NAGY,J.: Hungarian Patent 178 773 /1978/
(35) KNOLL,J., Physiology and Behaviour, 23 497 /1979/

THIN-LAYER CHROMATOGRAPHY

Proc. of the Symp. on Advances in Liquid Chromatogr., Szeged, Hungary, 1982

DETERMINATION OF BETA-BLOCKING AGENTS IN HUMAN URINE BY THIN-LAYER AND GAS CHROMATOGRAPHY

J. PUCSOK, I. HOLLÓSI

Research Laboratory of the National Institute for Medicine of Physical Education and Sports, Budapest, Hungary

Adrenergic beta receptor blockers are the therapy of choice for hyperkinetic heart syndrome /increased symphathoadrenergic activity/.Thus, there is an increasing interest in their use in competitive sport. The determination of beta-blockers in urine is important for understanding the mechanism of action of these drugs and in the future the methods of determination will possibly be very useful in doping control,too.

Tobanum hydrochloride /1-tert.-butyl-amino/-3-/2,5--dichlorophenoxy/-propanol-2 hydrochloride has antihypertensive, antiarrythmic and anticonvulsant activity. Our aim was to develop a simple and rapid qualitative thin-layer chromatographic method and a gas-liquid chromatographic assay for the quantification of some adrenergic beta-receptor blockers in human urine.

For thin-layer chromatographic analysis the urine samples /4 ml/ were made alkaline and extracted with etherdichloromethane 4:1 / 4 ml/ without enzymatic hydrolysis. After evaporation of the solvent the residue was chromatographed using pre-coated TLC plates, SILGUR 25 UV_{254} and Kieselgel G, respectively. The developing agent was ethylacetate-methanol, the visualising reagents were formaldehyde -sulphuric acid and Draggendorf reagent.

The R_F values of the compounds studied and their colors under UV light and after treating with visualising reagent are given in Table I. The minimum detectability of each drug is given in the same Table.

For gas chromatographic analysis the urine samples after enzymatic hydrolysis were made alkaline and extracted with ether--dichlormethane using Propranolol as internal standard.

TABLE I.

R_F values, characteristic colours and minimum detectability in urine Solvent system: ethylacetate-methanol-ammonia /40:5:5/

Compound	R_F Silgur 25	R_F Kieselgel G	UV_{254}	Colour with formaldehyde sulf.acid.	Colour with Draggendorf reagent	Minimum detectability in urine mg/l
Tobanum	0,86	0,62	+	--	red brown	0,5
Propranolol	0,69	0,39	+	green	red brown	0,5
Oxprenolol	0,78	0,48	+	purple	brown	0,5
Pindolol	0,72	0,41	+	brown	brown	1,0

The ether-dichloromethane 4:1 mixture was found to be the most suitable as it extracted the least number of endogeneous compounds and it gave adequate recoveries for both Tobanum and Propanolol. The recoveries were 65 % from urine. The physicochemical properties of Propanolol are closely related to those of Tobanum, therefore the use of this compound as an internal standard was found to be satisfactory. Known quantities of Tobanumhydrochloride were added to blank serum samples. Calibration curves were constructed by plotting the peak area ratios between Tobanumhydrochloride and the internal standard versus the amount of Tobanumhydrochloride added. Subsequently both compounds were back-extracted into hydrochloric acid and then into ether-dichloromethane again. After evaporation of the organic phase under nitrogen atmosphere the compounds were derivatized with N-methyl-bis-trifluoroacetamide /Fig.1./. These derivatives were analysed on a gas-liquid chromatograph /Hewlett Packard 7620A/ equipped with a ^{63}Ni-electron-capture detector and a glass column /1.20 m x 4 mm i.d./ packed with 3,8 % UCW-982 on Chromosorb W /100-120 mesh/ and connected to a recorder with a scale range of 1 mV. The carrier gas was argonmethane 95:5 at a flow rate of 60 ml/min. Temperatures: injection port 230 C^o, column 180 C^o, detector 250 C^o /Fig.2/. A gasliquid chromatograph coupled to a mass spectrometer /HP 5992/ was used for the characterization of Tobanum and Propranolol /Fig.3., Fig.4./. GLC peaks of Tobanum-bis-trifluoroacetate should give a molecular ion at m/z 483.

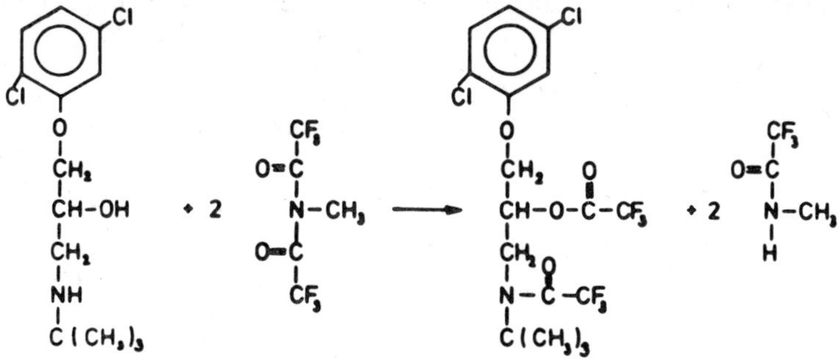

Figure 1. Trifluoracetylation of Tobanum

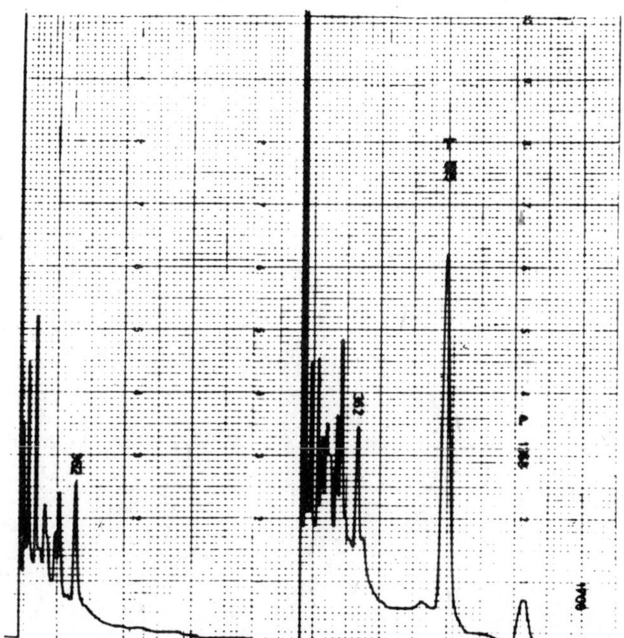

Figure 2. Gas chromatogram of A. blank urine extract B. urine extract obtained from a volunteer 8 hours after administration of 10 mg Tobanum.

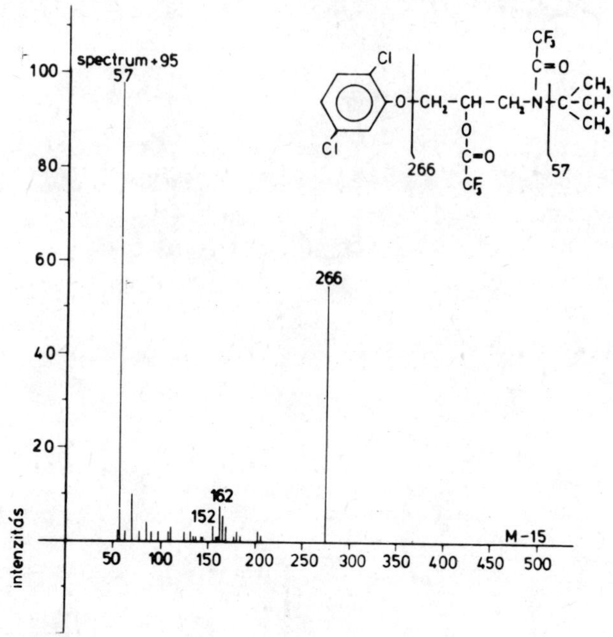

Figure 3. Low-resolution mass spectrum of Tobanum-bis-trifluoracetate.

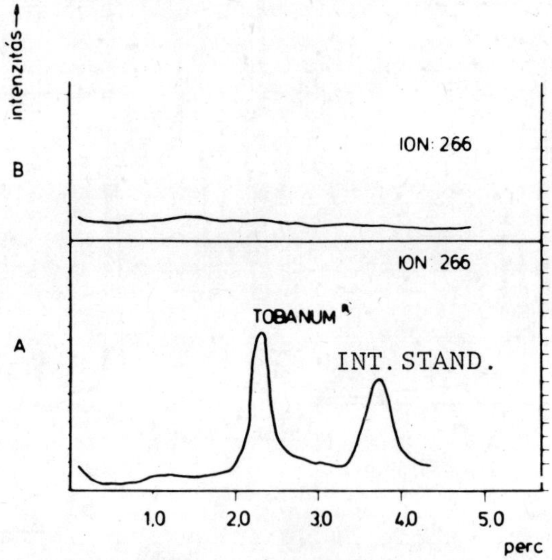

Figure 4. Single-ion /m/z 266/ recordings of urine extracts A. urine from a volunteer treated with Tobanum B.blank urine.

The mass spectrum exhibited only an ion at m/z corresponding to the loss for a methyl group from the molecular ion. The ion at m/z 162 is formed by loss of the side chain after H-arrangement and the fragment ions at m/z 266 and m/z 152 arise from cleavage of the side chain at various positions. The base peak at m/z 57 corresponds to the terc-butyl group. The intense fragment ion at m/z 266 is characteristic of aryloxy beta-blocking drugs.

The mass spectrometric information was obtained in the electronimpact mode under the following conditions: ionization beam energy 70 eV, electron multiplier 2800 V. The glass column /3 ft x 2 mm i.d./ was packed with 2 % OV-101 on Gas-Chrom Q /100-120 mesh/ and operated at 170 C^{o} with a helium gas flow rate of 25 ml/min.

Our methods are relatively simple and rapid. The minimum detectability of the beta-blockers in urine by thin-layer chromatography is 0,5 mg/l. The gas chromatographic method can be used in the concentration range 10-2000 ng/ml for routine determination as well. The methods have now been applied to urine specimens from healthy volunteers who have taken these drugs orally. 24 hours after the oral administration the drugs are still detectable.

REFERENCES

(1) SOLTI,F. and CZAKÓ,E.: Antiarrhythmic properties of Tobanum[R]. A new beta-blocking agent
Advances in Pharmacological Research and Practice
Vol.I. Pharmacological Control of Heart and Circulation
277-281, 1979

(2) KECSKEMÉTI,V.: Effects of Tobanum /GYKI-41099/ on cardiac transmembrane potentials
Advances in Pharmacological Research and Practice
Vol. I. 263-266, 1979

(3) ELEKES,I.,TOMORI,É.,PÁTFALUSI,M.,PAST,T.: Human pharmacokinetic study on Tobanum[R].
Advances in Pharmacological Research and Practice
Vol.I. 267-272, 1979

(4) THOMAS WALLEE: GLC determination of Propranolol, other beta-blocking drugs and metabolites in biological fluids and tissues
J.Pharm.63, 1885 /1974/

(5) DEGEN,P.H. and RIESS,W.: Simplified method for the determination of Oxprenolol and other beta-receptor blocking agents in biological fluids by gas-liquid chromatography
J.Chromatogr. 121 72 /1976/

(6) GUERRET,M.,LAVENE,D. and KIECHEL,J.R.,Determination of Pindolol in biological fluids by electron-capture GLC J.Pharm. 69 1191 /1980/

(7) JACK,D.B., DEAN,S.,KENDALL,M.J. and LEUGHER,S.,Detection of some antihypertensive drugs and their metabolites in urine by thinlayer chromatography.
I.Five Commonly used beta blockers and hydralazine
J.Chromatogr. 187 277 /1980/

(8) PETERSON ,J.W., CONOLLY,M.E.,DOLLERY,D.T.,The pharmacodynamics and Metabolism of Propranolol in Man Pharmacologia, Clinica 2, 127 /1970/

(9) DONIKE,M., J.Chromatogr. 43 273-279, /1973/

(10) TOMORI,E.,ELEKES,I. ,Specific and sensitive method for the determination of Tobanum using gaschromatography with electron capture and mass fragmentographic detection
J.Chromatogr. 204 355 /1981/

(11) THOMAS WALLE, MORRISON,J.,WELLE,K., Simultaneous determination of propranolol and 4-hydroxypropranolol in plasma by mass fragmentography
J.Chromatogr. 114 351 /1975/

Proc. of the Symp. on Advances in Liquid Chromatogr., Szeged, Hungary, 1982

CHEMICAL STRUCTURE AND LIQUID CHROMATOGRAPHIC BEHAVIOUR AMONG NITROGEN-BRIDGED COMPOUNDS

A. SHALABY, ZS. BUDVÁRI-BÁRÁNY, K. HANKÓ-NOVÁK, GY. SZÁSZ, I. HERMECZ*

Institute for Pharmaceutical Chemistry, Semmelweis Medical University, Budapest
*Chinoin Pharmaceutical Works, Budapest, Hungary

One of our working teams is dealing with the synthesis of biologically active compounds. In the scope of this systematic research activity during the last years a rather large number of heterocyclic compounds mostly nitrogen-bridged derivatives were prepared (1,2). Elucidation of relationships between chemical structure and biological activity within this type of compounds seemed to be a rather interesting and promising task. It is well known that biological effect and certain physical properties are in many cases closely related. The partition coefficient in octylalcohol-water system is generally regarded as a classical parameter in this field. One can meet numerous initiatives trying to substitute the partition coefficient by other physical parameters. Recent papers have indicated that thin-layer chromatographic R_M value has proved as to be an adequate parameter to partition coefficient. Papp et al. (3) reported on linear correlation between gas chromatographic retention index and partition coefficient just in the same circle of the compounds we are dealing with in this paper.

Regarding its simplicity, effectiveness and overall applicability, velocity, it seemed to be worthwhile to pay special attention to the retention data /i.e. log k' values/ which can be obtained by HPLC. However, one must take into consideration that correlation between liquid chromatographic retention and partition coefficient needs a chromatographic process exclusively governed by liquid/liquid distribution forces. Certain authors have reported difficulties just at the circle of aminelike compounds (4). We have also similar experiences and in this report

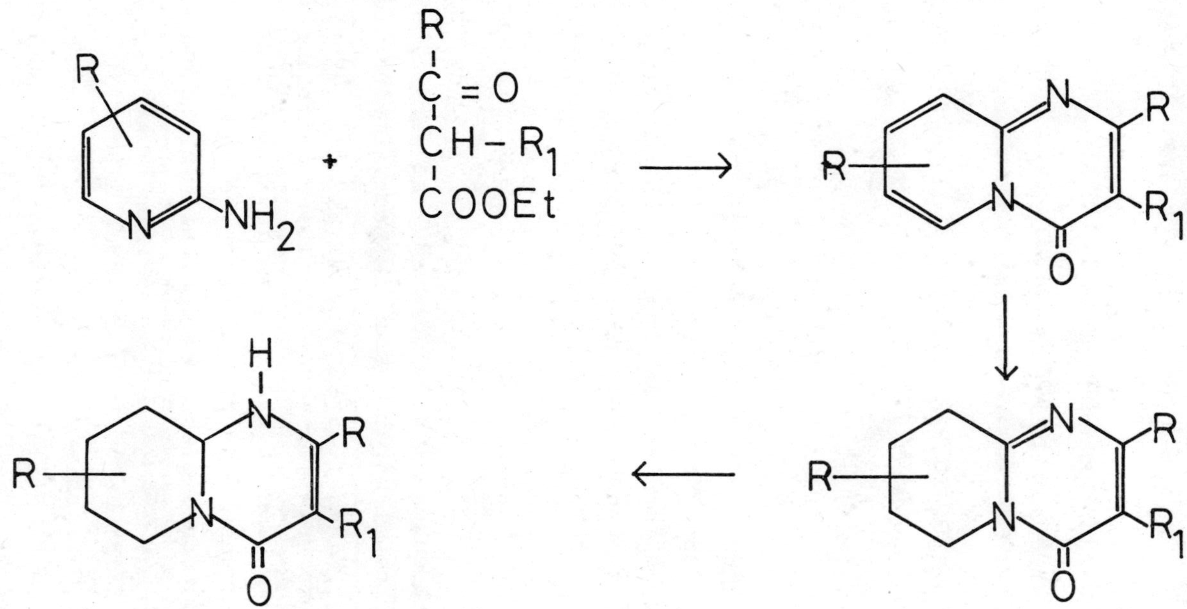

Figure 1. The scheme of synthesis of pyridopyrimidine derivatives

on first results with certain types of nitrogen bridged compounds are to be published.

One part of our model substances comes from the type of pyridopyrimidine and were prepared according to the scheme depicted in Fig.1. The synthesis was based on the condensation of 2-amino-pyridine with a suitable ketoacid derivative. Figure 2. shows an other synthesis of chinazolon derivatives with three cycles and also with five rings. In the mentioned way, the total synthesis of an alkaloid namely rutecarpine and its artificial derivatives could be performed with a rather good yield (5). In this paper those results obtained with the derivatives of the two and three cyclic compounds you can see in Fig.3. will be given. Among the studied derivatives these are compounds with saturated aromatic rings which are in every case condensed with a central 4-pyrimidion part. It is also to be noted that the third ring is a five or six membered one. In Fig.4. the derivatives having unsaturated "A" ring can be seen. The behaviour for the different alkyl derivatives in the unsaturated two and three ring series and also certain compounds from the saturated series /Fig.5/ were studied.

In Fig.6 the partition coefficients and log P values (6) of the unsubstituted compounds of the structural types we have studied are presented . Observing these values an obvious conclusion emerges: both these types of compounds have a rather polar structure./ It may be of interest that the saturated derivatives proved to be significantly more polar than the others./ Anyway, the relatively high polarity of the compounds has made it justifiable to work with reversed phase during the HPLC-study.

In Fig.7 log k' values of fourteen unsaturated derivatives have been collected. In the first column the structure of compounds and in the next one the log k' values at different ratios of the mobile phase methanol-water are visible. The methanol concentration of the mixtures varied from 50 % to 90 %. In the last column the regression coefficients for the compounds are visible. The stationary phase was Lichrosorb C_{18}, particle size 7 mμ. One can see the parallel increase of log k' values with the increase of water concentration. It is also evident that there is a correlation between log k' and C_{MeOH} in the studied range.

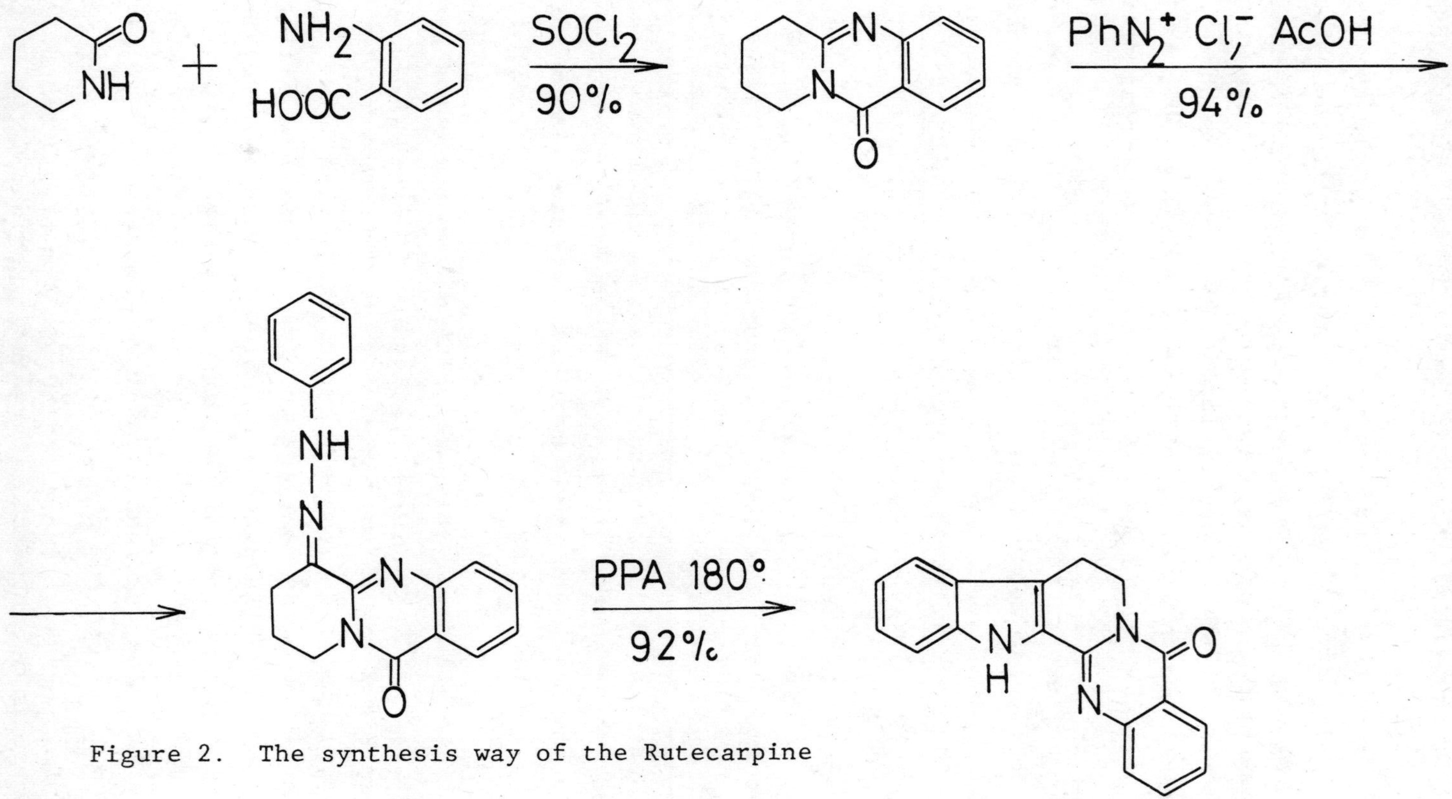

Figure 2. The synthesis way of the Rutecarpine

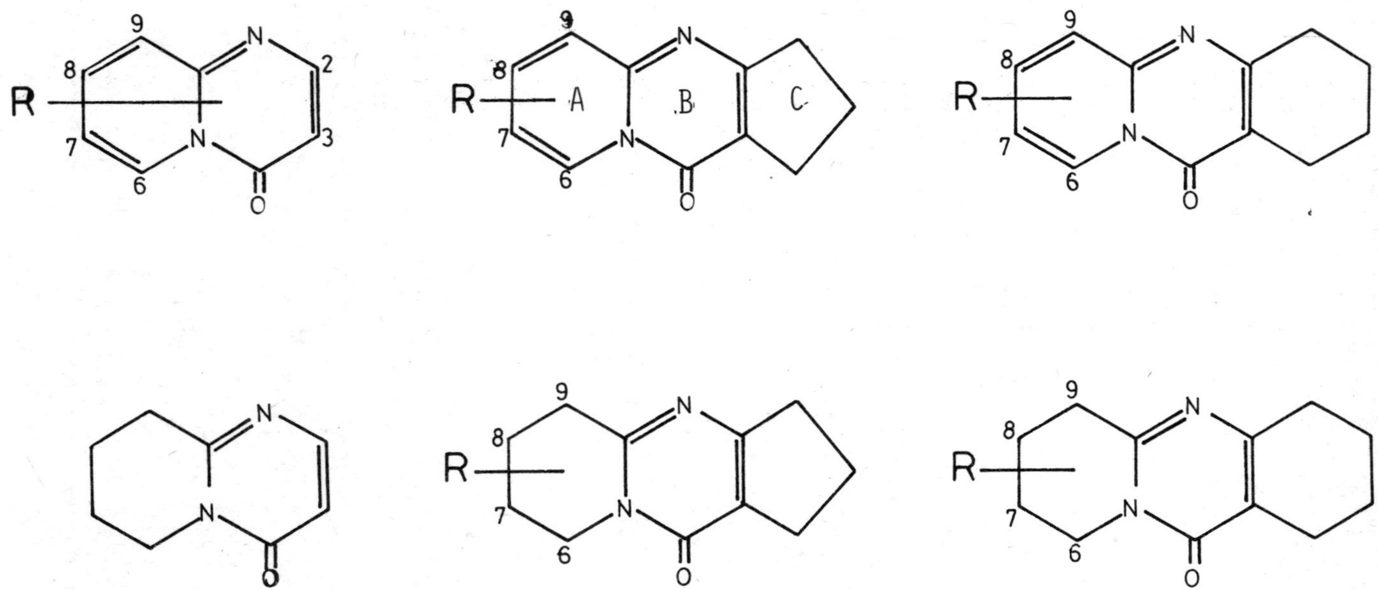

Figure 3. The investigated compounds with two and three cycles

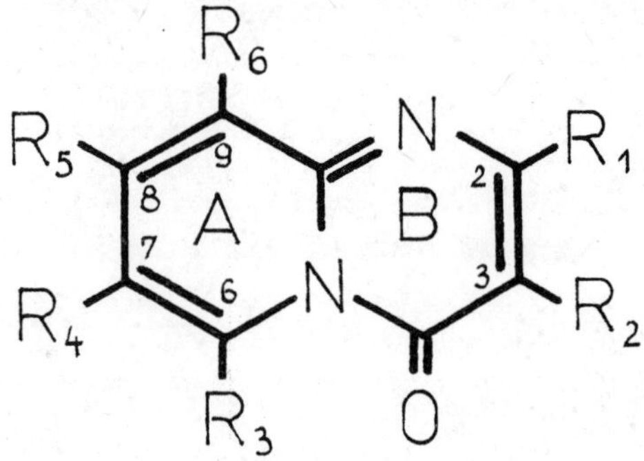

	C_2	C_3	C_6	C_7	C_8	C_9
1.	H	H	H	H	H	H
2.	CH_3	H	H	H	H	H
3.	H	CH_3	H	H	H	H
4.	H	H	CH_3	H	H	H
5.	H	H	H	CH_3	H	H
6.	H	H	H	H	CH_3	H
7.	H	H	H	H	H	CH_3
8.	CH_3	H	CH_3	H	H	H
9.	H	CH_3	CH_3	H	H	H
10.	CH_3	H	C_2H_5	H	H	H
11.	H	C_2H_5	CH_3	H	H	H
12.	CH_3	C_2H_5	CH_3	H	H	H
13.	C_2H_5	CH_3	CH_3	H	H	H
14.	C_3H_7	C_2H_5	CH_3	H	H	H

Figure 4. Structure of compounds with unsaturated "A"-ring

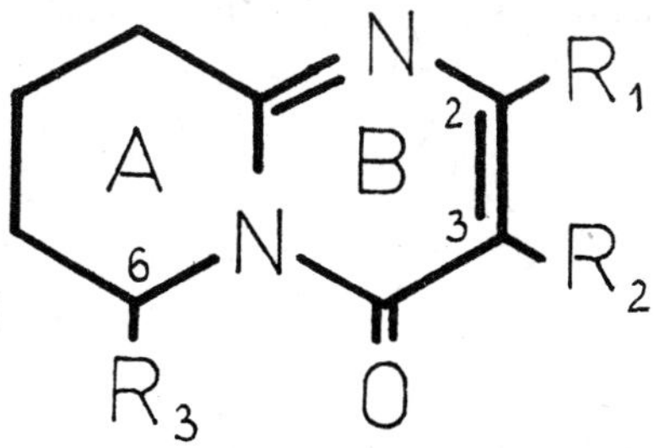

	C_2	C_3	C_6
1.	H	H	H
2.	H	H	CH_3
3.	CH_3	H	CH_3
4.	H	CH_3	CH_3
5.	CH_3	C_2H_5	CH_3

Figure 5. Structure of compounds with saturated "A"-ring

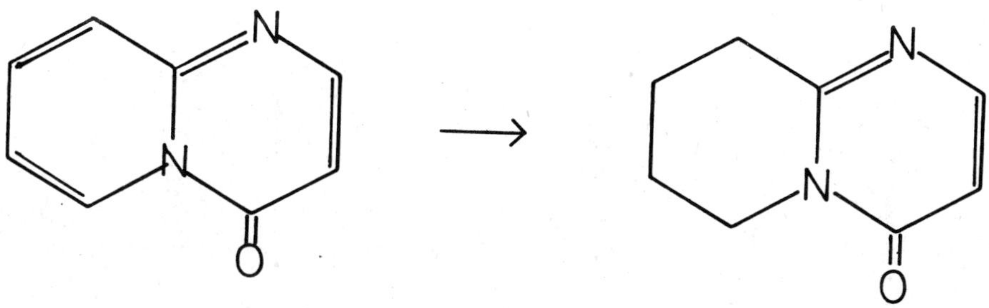

$\log P = 0{,}20$ $\quad$ $\log P = -0{,}14$

$P = 1{,}60$ $\quad$ $P = 0{,}72$

Figure 6.

Unsaturated compounds substituent	MeOH %					r = regression coeff. of log K'/MeOH conc.
	90	80	70	60	50	
	- 0,804	-0,756	-0,580	- 0,394	-0,252	0,978
C_2 – Me	- 0,638	- 0,677	-0,477	- 0,148	-0,035	0,971
C_3	- 0,585	- 0,518	-0,357	- 0,143	0,115	0,986
C_6	- 0,618	- 0,572	-0,358	0,070	—	0,936
C_7	- 0,650	- 0,649	-0,413	- 0,143	0,086	0,981
C_8	- 0,809	- 0,669	-0,477	- 0,120	0,107	0,988
C_9	- 0,723	- 0,602	- 0,375	- 0,164	0,065	0,996
C_{2-6} -di-Me	- 0,509	- 0,394	- 0,188	—	—	0,987
C_{3-6}	- 0,397	- 0,217	0,065	—	—	0,992
C_2-Me, C_6-Et	- 0,538	- 0,301	0,00	—	—	0,998
C_3-Et, C_6-Me	- 0,276	- 0,082	0,147	—	—	0,999
C_2-Et, C_{3-6}-di-Me	- 0,207	0,023	0,301	—	—	0,998
C_{2-6}di-Me, C_3-Et	- 0,174	0,316	0,399	—	—	0,925
C_2-Pr, C_3-E, C_6-Me	0,000	0,265	0,671	—	—	0,993

Figure 7. The log k' values in different MeOH-H_2O solvent mixtures on C_{18}-reversed phase
←

On the contrary the picture is not so simple for saturated compounds /Fig.8/. The value of the correlation coefficient is acceptable only for 2 or 3 compounds from the 5. The main cause for this is to be found in the different stereochemical relations of the two rings within the saturated and unsaturated types. It may be seen that in contrast to the completely planar unsaturated ring-system the cycloaliphatic "A"-ring closes a certain angle with the B-ring and the substituents situated nearer to the N^1 and C^4 atoms there by having a larger influence on the ability of these atoms to develop interactions with the components of stationary and mobile phases.

Figure 9. shows the overall correlation coefficients calculated for the different ratio of methanol-water. The coefficients are even in the best case rather low.

Figure 10. represents the relationship between log P log k' values including only the individual data of the fourteen unsaturated compounds. It is obvious that the values of four compounds are extremely deviating and this is the main cause that overall r-value is still rather low. However, calculating only with the data of the other ten compounds a good correlation is manifested and the regression coefficient for these ten compounds approaches the unit, being 0,98. It is very instructive to observe the structure of the four compounds showing extremely deviating log k' values and causing the poor correlation. At these compounds the substituents are in ortho-position to the distinguished N^1 and C^4 atoms which are mainly responsible for the HPLC behaviour of the molecules. It is reasonable to imagine that the ortho-effect of the mentioned groups stands out to a different extent against octylalcohol and methylalcohol during the separatory funnel-method and the PHLC method respectively.

Saturated compounds substituent	Me OH %					regression coeff. of logK/MeOH conc.
	90	80	70	60	50	
	0,678	0,845	0,544	0,340	0,227	0,937
C_6-Me	0,809	0,794	0,469	0,251	0,103	0,984
C_{2-6}-di-Me	0,721	0,027	0,405	0,133	0,042	0,837
C_{3-6}-di-Me	0,684	0,602	0,271	0,049	0 167	0,978
C_{2-6}-di-Me, C_3-Et	0,456	0,301	0,100	0,472	0,615	0,985

Figure 8. The log k values in different MeOH-H_2O solvent mixtures on C_{18} reversed phase

MeOH-H_2O conc. %	Regression coefficient of compounds with unsaturated „A"- ring	
50	—	
60	0,7586	n = 7
70	0,8614	n = 14
80	0,9539	
90	0,9439	

Figure 9. Relationship between log k' and log P values

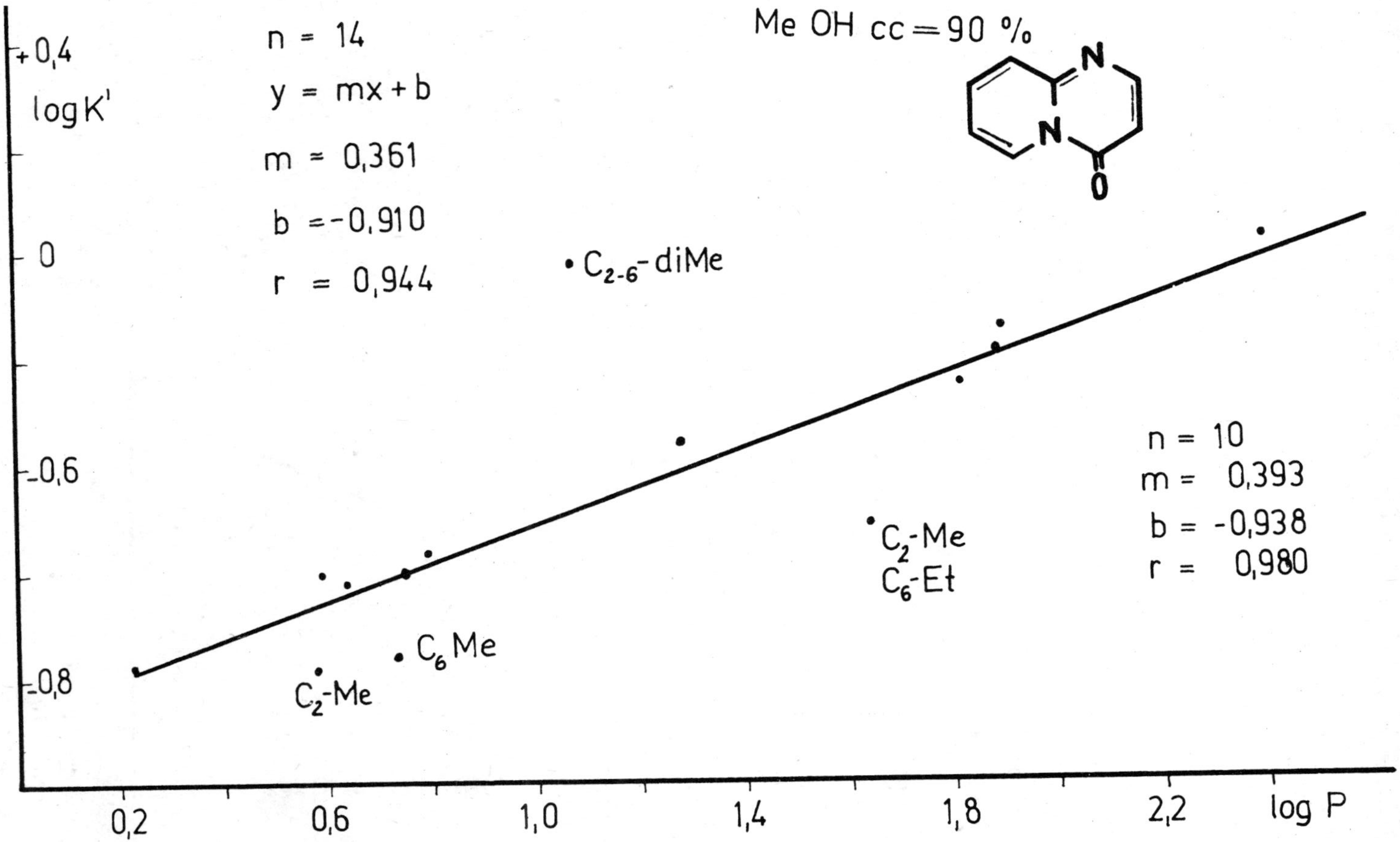

Figure 10. Linear regression of log P/log k of alkyl derivatives

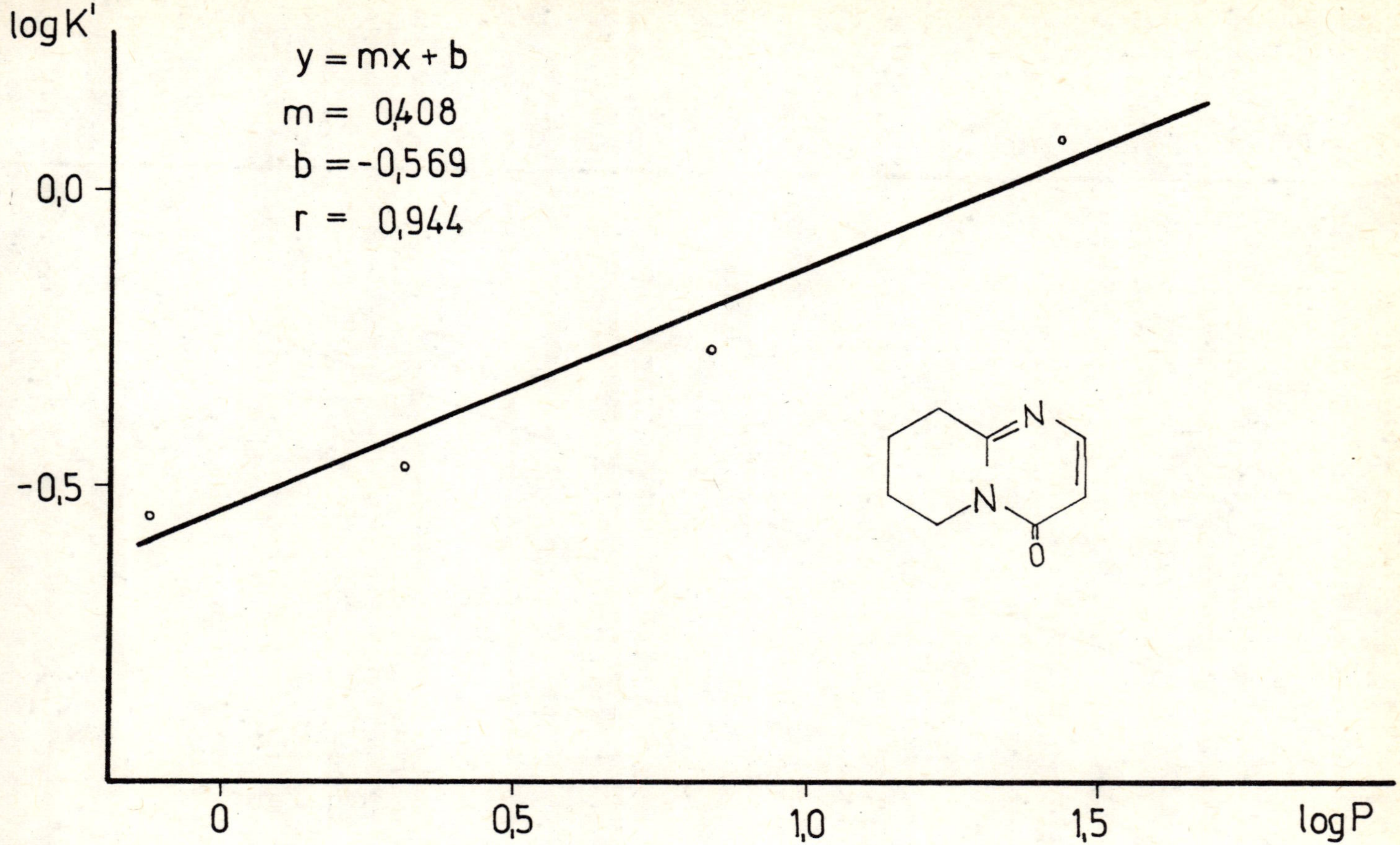

Figure 11. Linear regression of log P/log k of saturated alkyl derivatives MeOH cc=70%

Compounds	MeOH %				
	90	80	70	60	50
C_2 - methyl	0,166	0,079	0,103	0,246	0,287
C_3 -"-	0,219	0,238	0,223	0,251	0,368
C_6 -"-	0,186	0,184	0,222	0,324	—
C_7 -"-	0,155	0,106	0,167	0,251	0,338
C_8 -"-	0,004	0,087	0,103	0,274	0,359
C_9 -"-	0,081	0,154	0,205	0,230	0,317

Figure 12. Δ log K' values of monomethyl compounds
$\Delta \log K' = \log K'_{R-X} - \log K'_{R-H}$

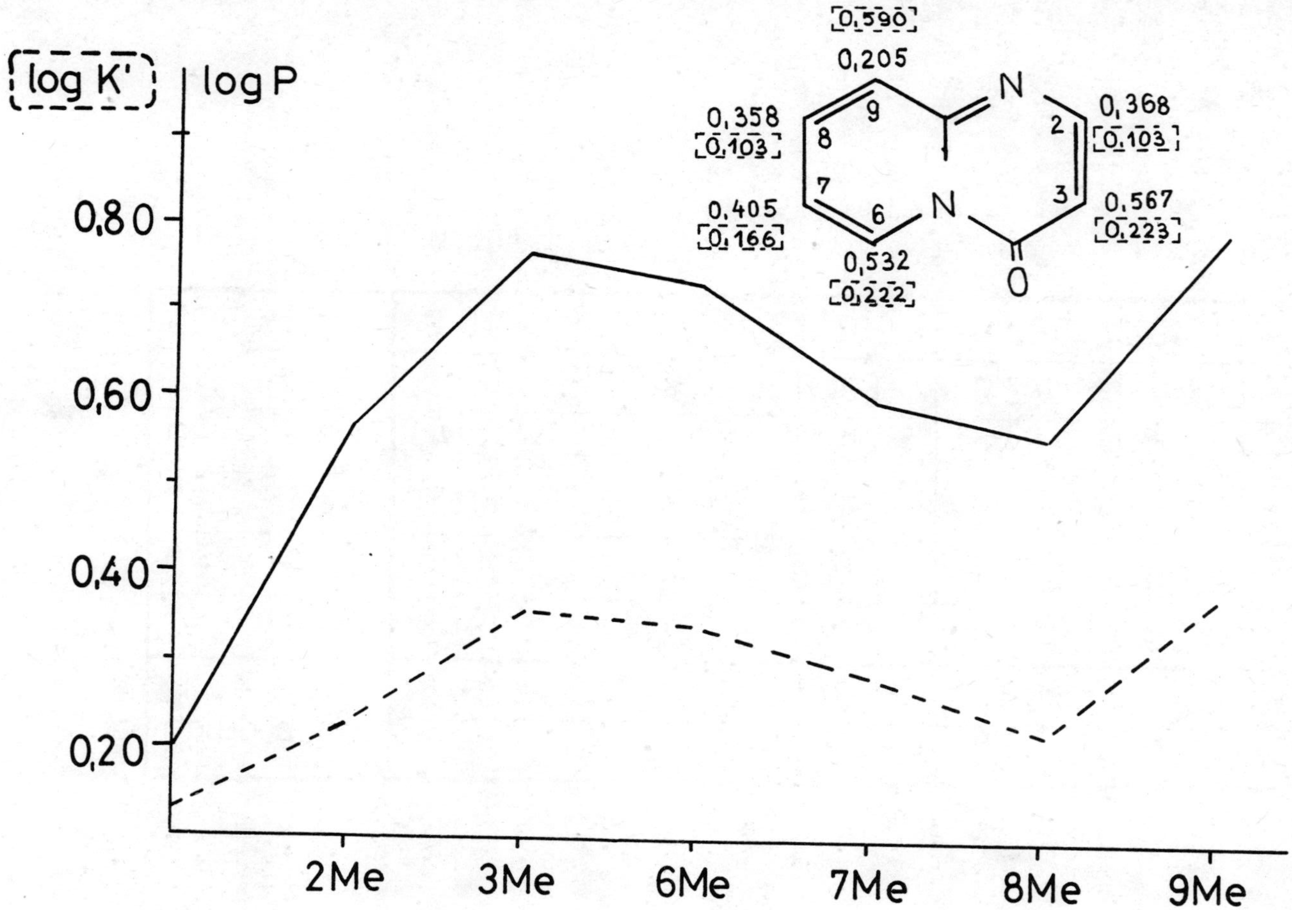

Figure 13. Relationship between log P and log K' values

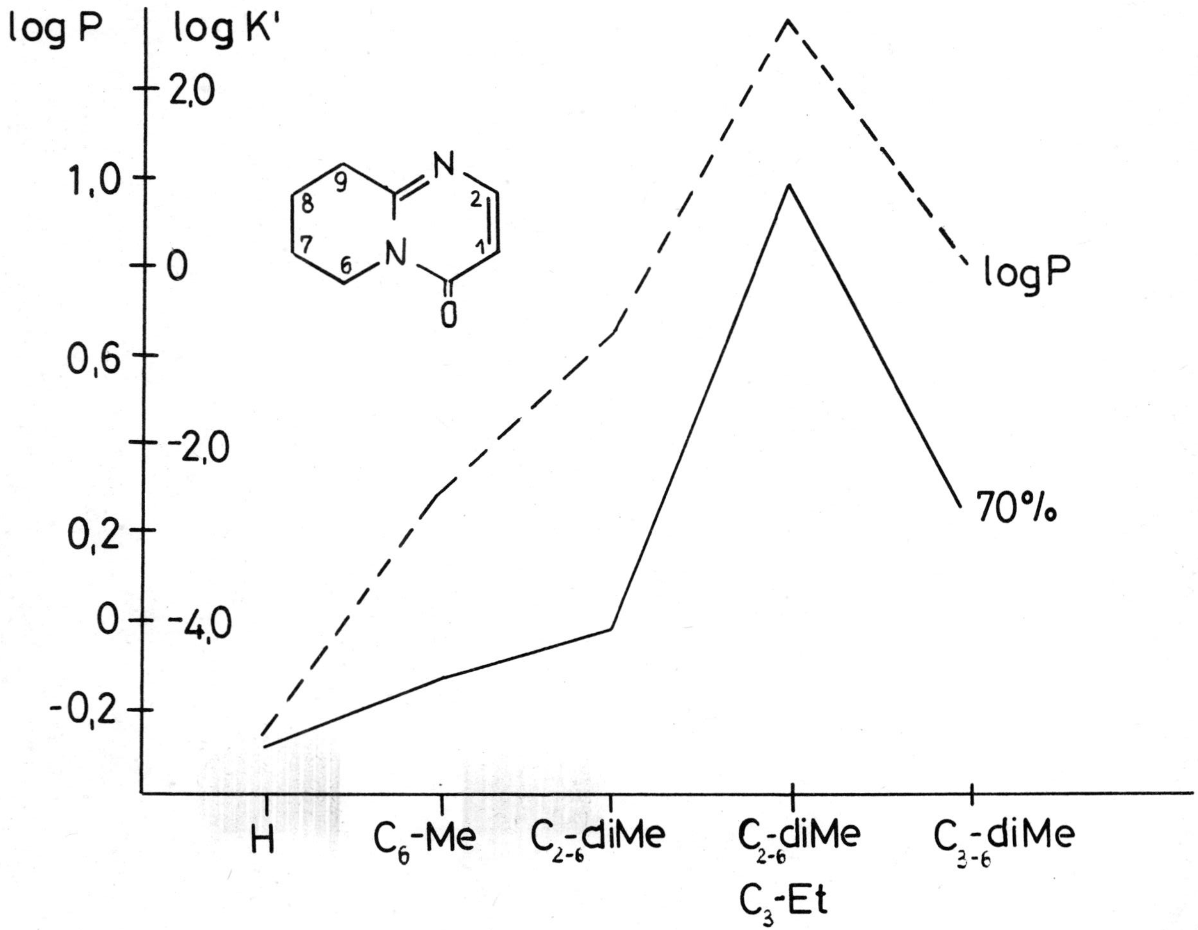

Figure 14. Relationship between log P and log K' values

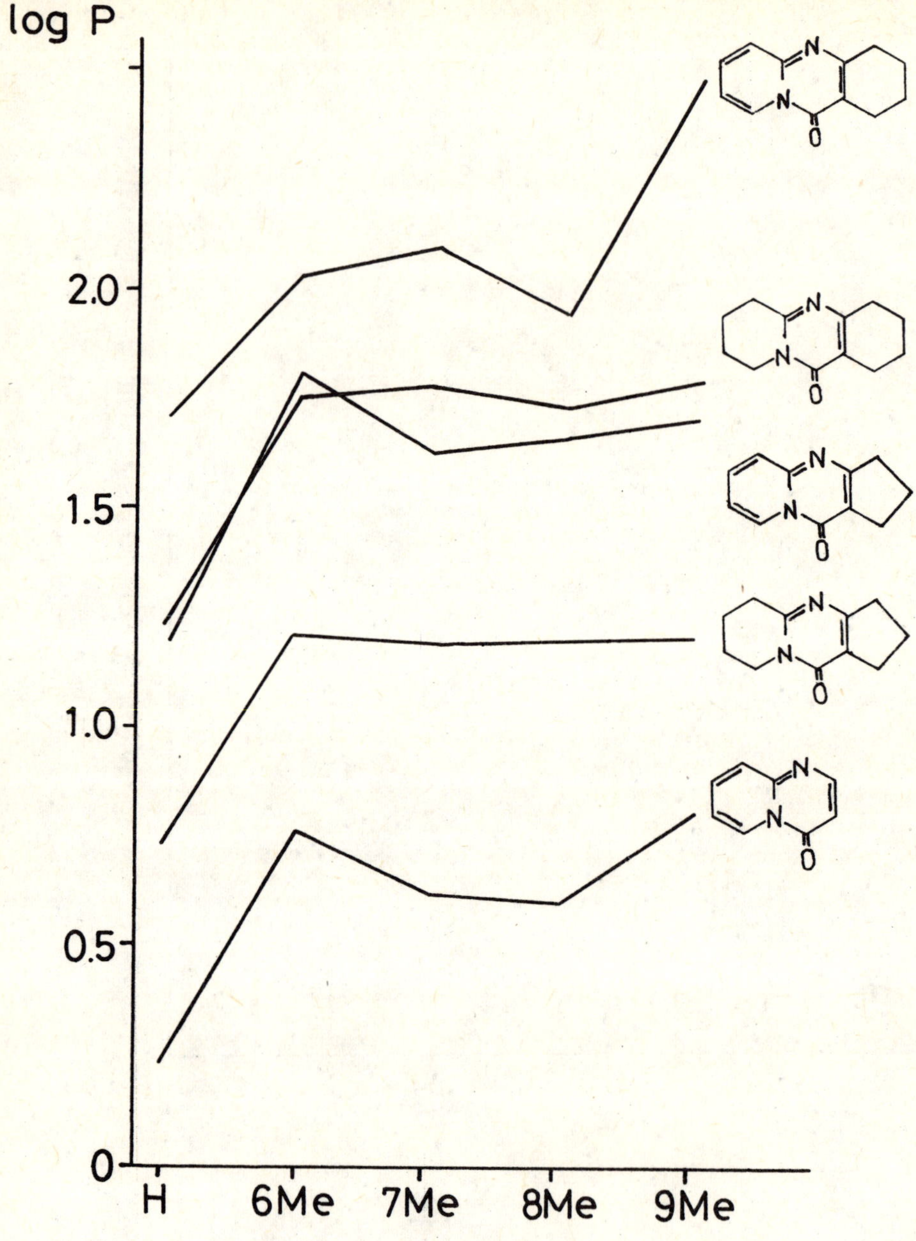

Figure 15. Log P values of threecyclic monomethyl compounds

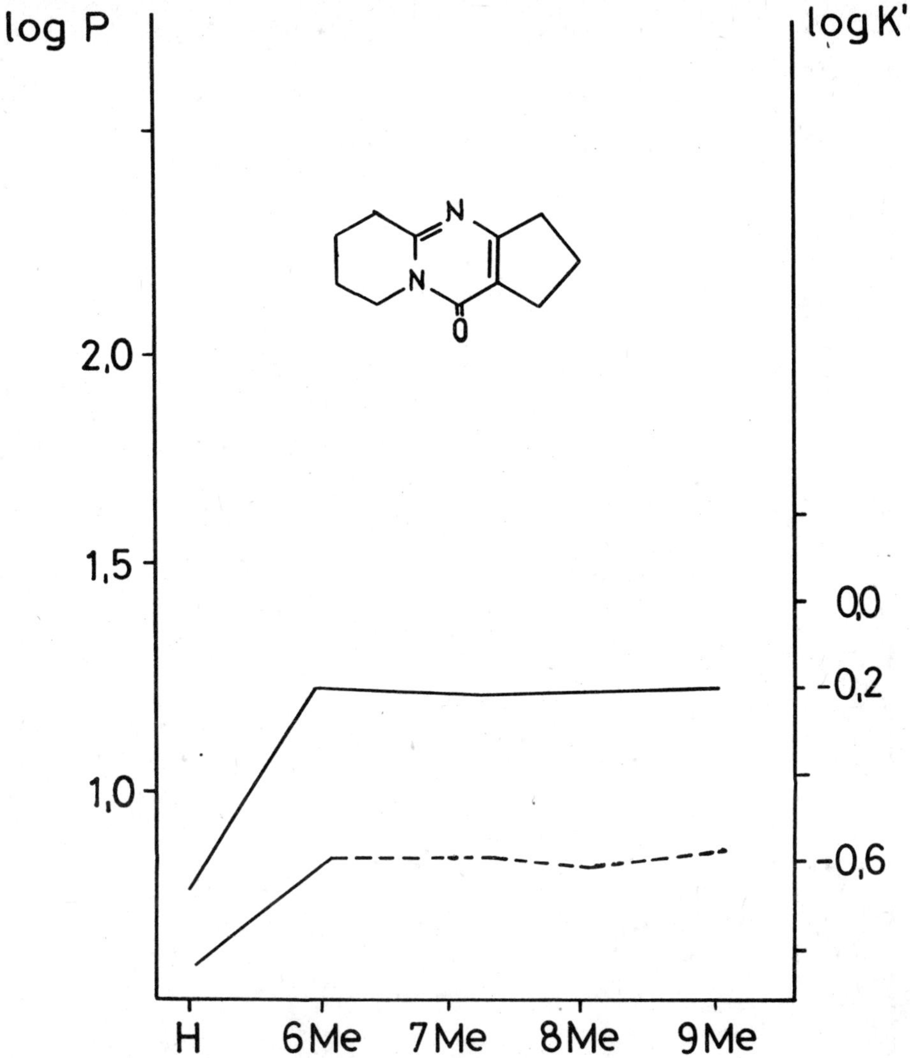

Figure 16a. Comparison of log P and log k' curves of threeecyclic monomethyl compound.

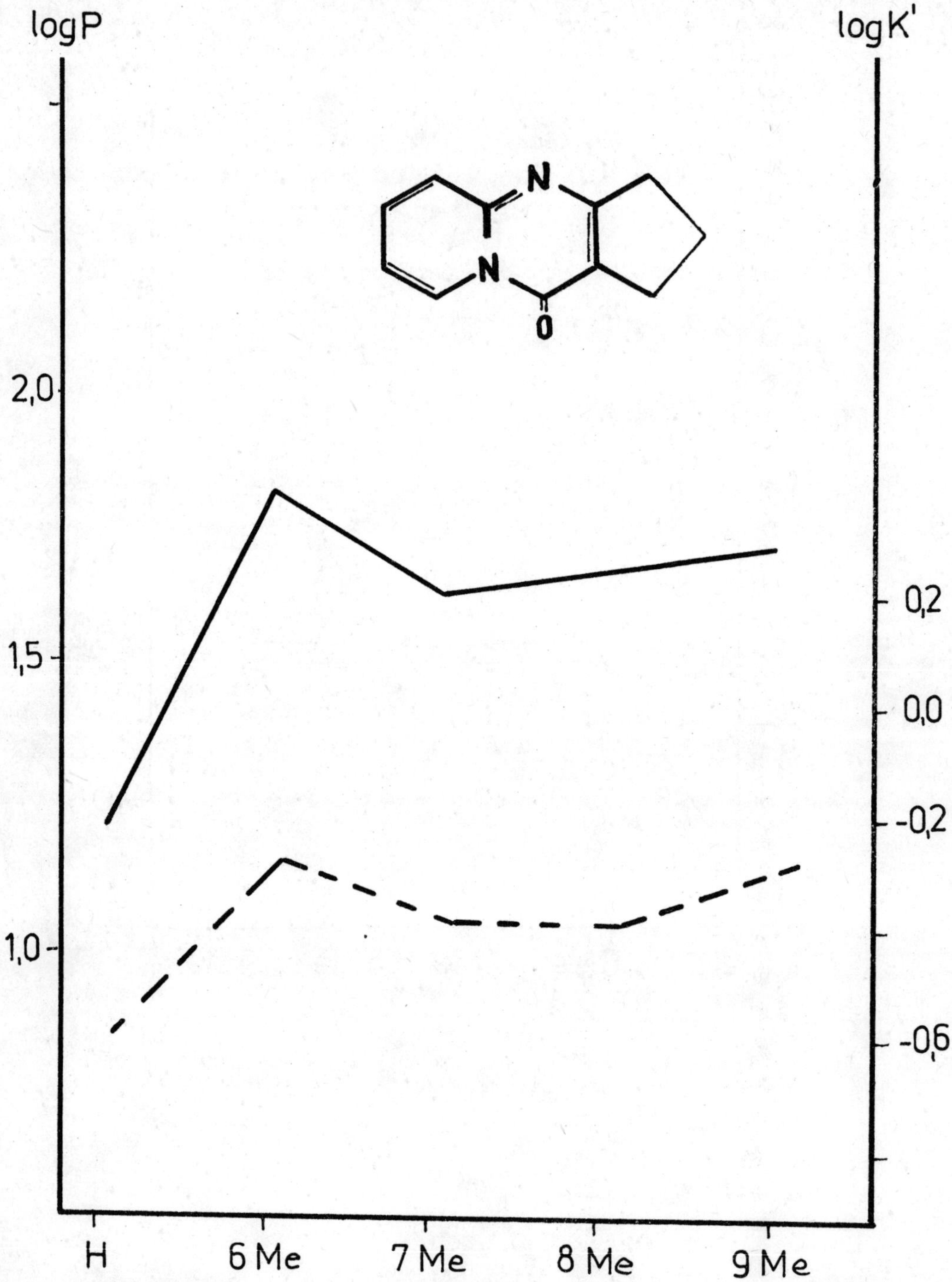

Figure 16b.

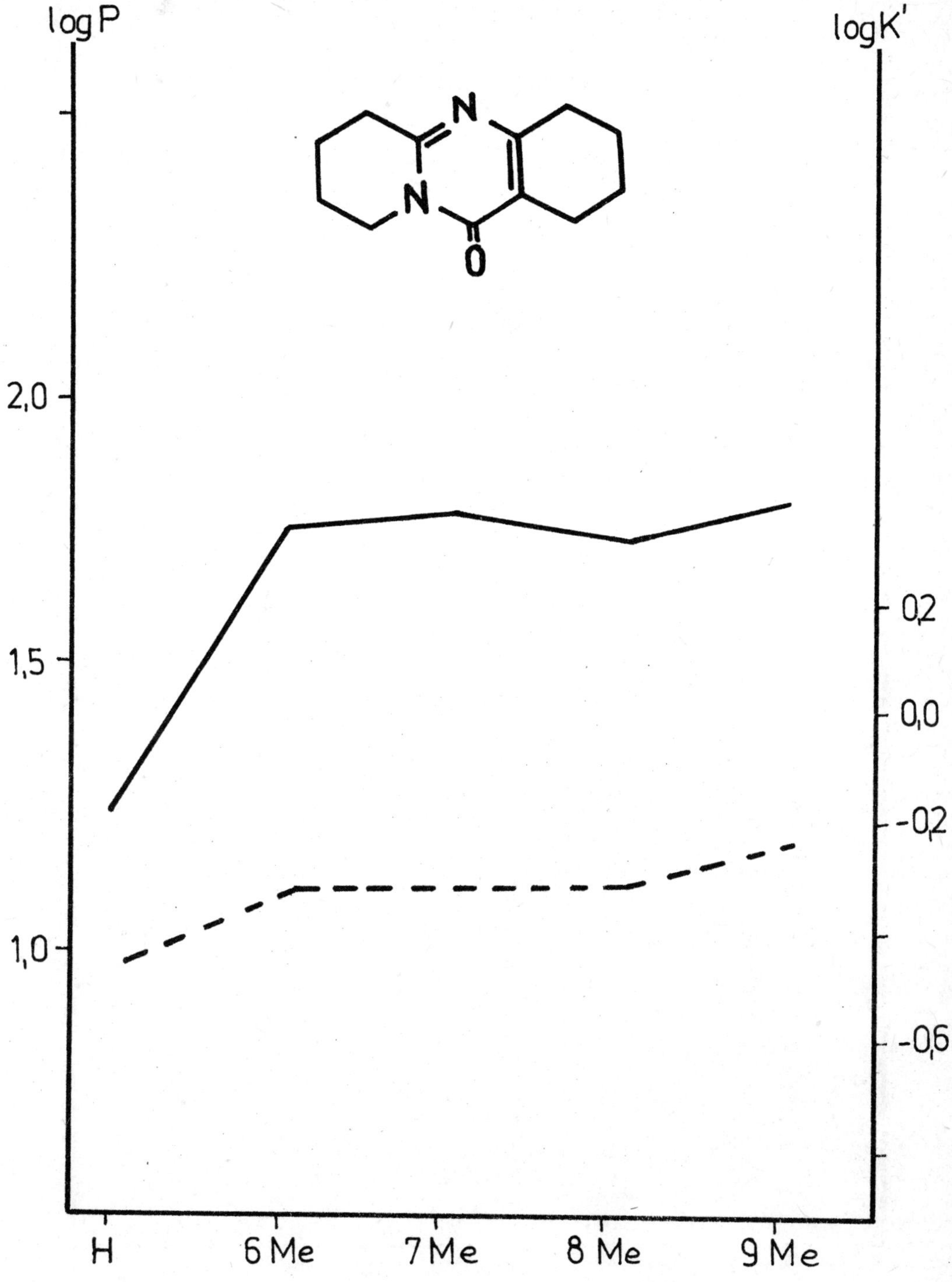

Figure 16c.

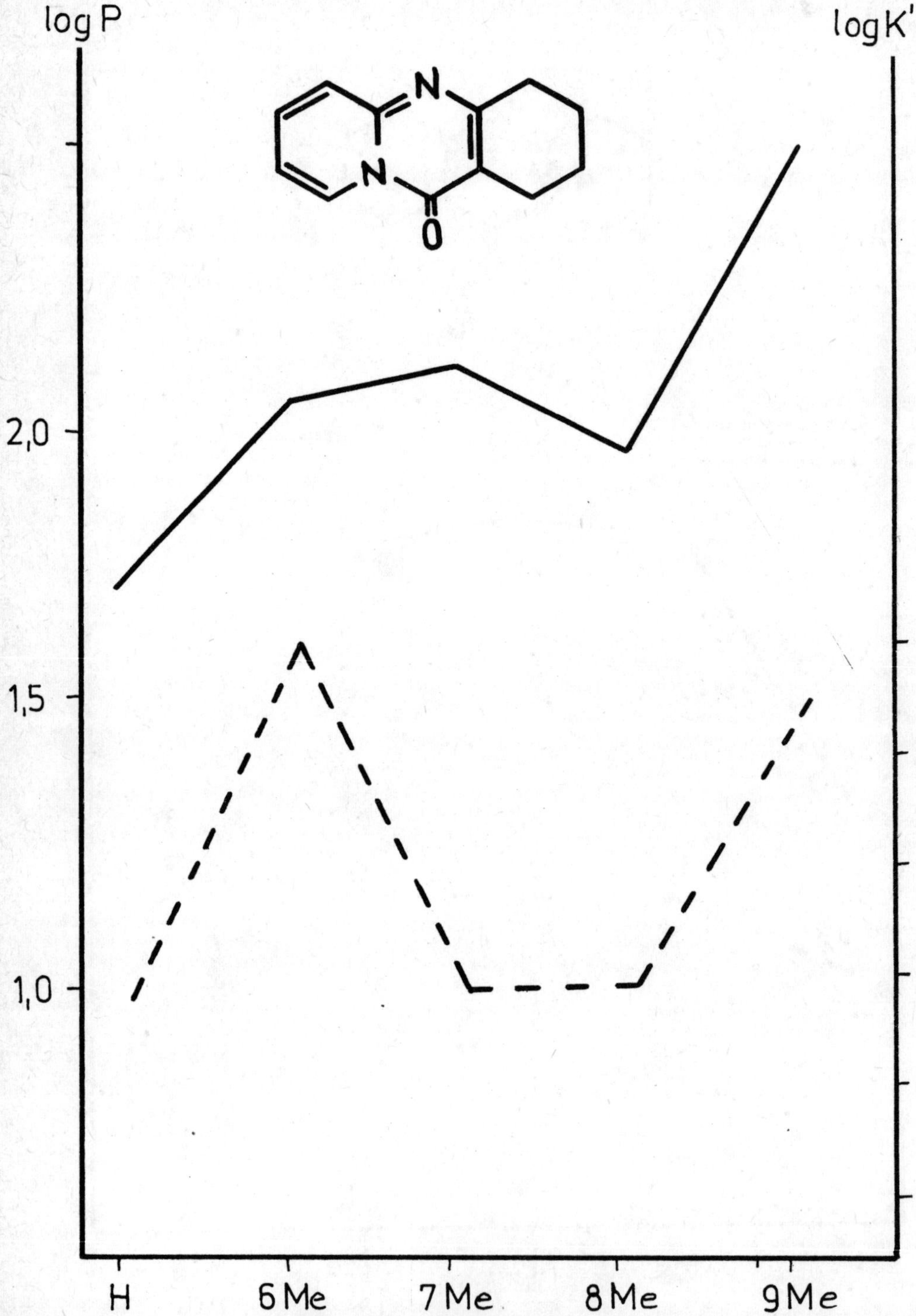

Figure 16d.

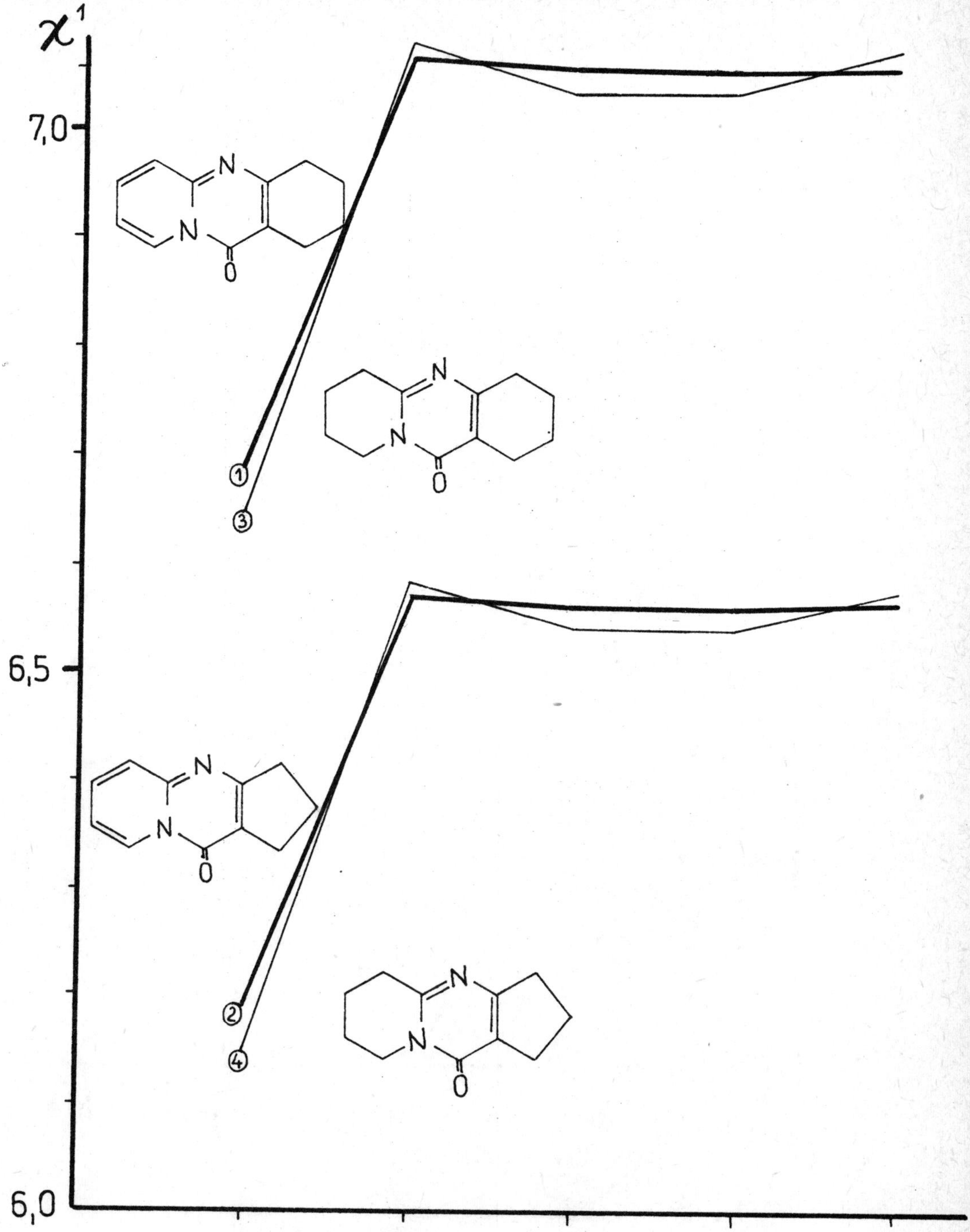

Figure 17. The χ^1 values related to methyl substitution

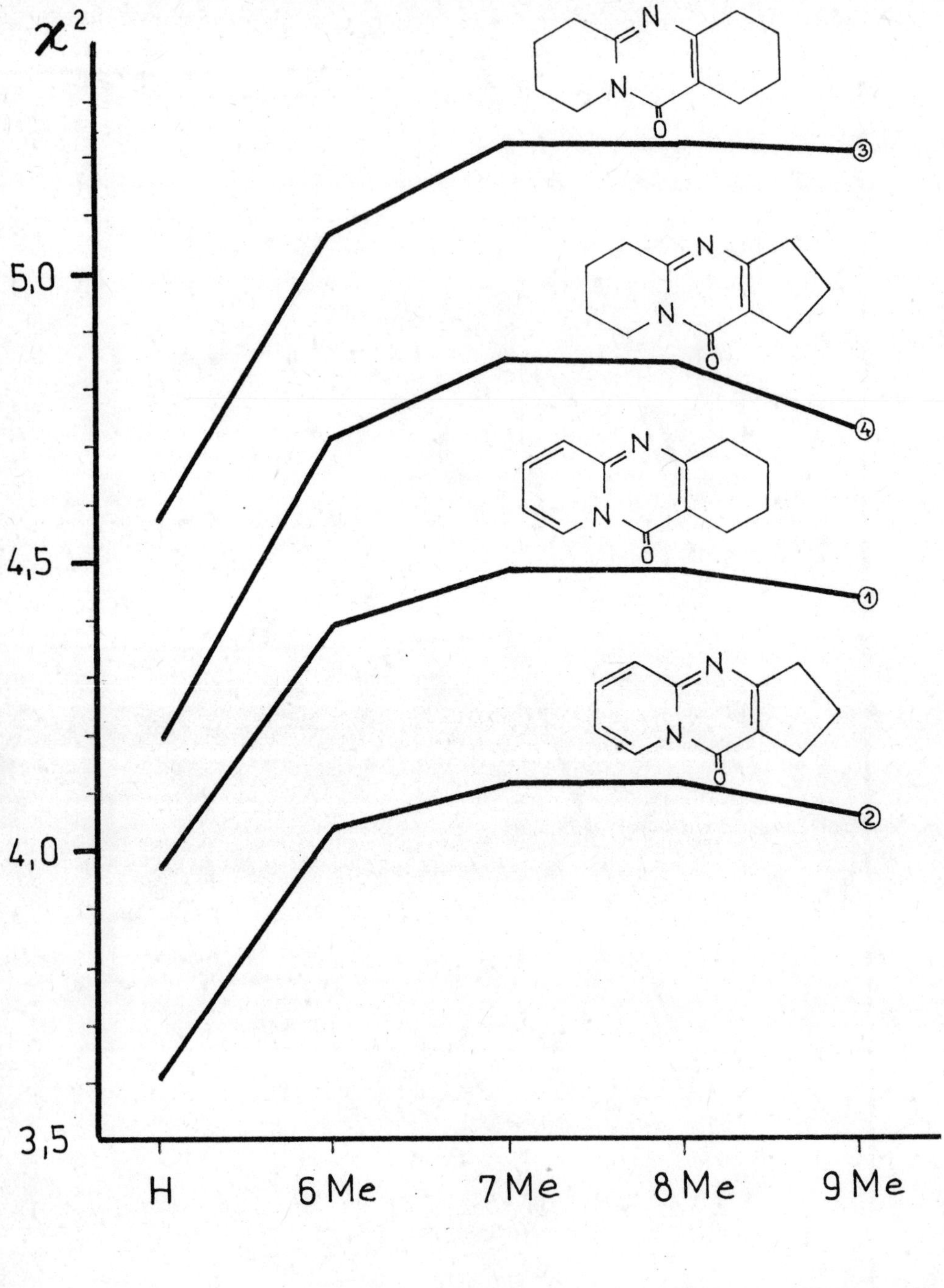

Figure 18. The χ^2 values related to methyl substitution

The correlation between log P and log k' among saturated derivatives /Fig.11/ is homogenously moderatedas to be expected as a consequence of the stronger sterical influence of the substituents on cycloaliphatic ring.

Fig.12 shows the Δ log k' values of the monomethyl-derivatives. By definition Δ log k' is a value which corresponds to the π-value of Hansch. As it might be expected the value of Δ log k' i.e. the effect of methyl-substitution increases together with the water content of the mobile phase.

Fig.13. shows curves based on the log P and log k' values of the monomethyl-derivatives in the unsaturated series. The character /shape of the curves of the pairs is rather identical. This means that the interactions within the separatory funnel and within the HPLC system are qualitatively identical but there is a difference in the extent or in the strength of the interaction. The hydrophobic effect of C^9-methyl group proved to be the strongest by both of the methods. The picture is similar in the case of saturated monomethyl-derivatives /Fig.14/.

Figs. 15-16. summarise the experiences with the methyl derivatives of three cycles. The increase of log P values /Fig.15/ is clearly related to increase in hydrophobicity /i.e.: is related to the increase of unsaturated character of the ring system/. From this point of view the C_9-methyl group has a distinguished importance again. The effect on hydrophobicity is almost the same on behalf of the C_6-C_7- and C_8-methylgroups. Comparing the log P and log k' curves of the different methyl derivatives in the case of the four threecyclic types the shape of the curve-pairs may be estimated as identical.

Fig.17. shows the first order values of molecular connectivity /X^1/. It is interesting to mention that values of ring system 6-6-5 and the type 6-6-6 coincide with each other, X^1 does not differentiate these ring-systems.

The second order values of molecular connectivity /X^2/ of the four types of threecyclic methyl-derivatives may be seen in Fig.18. In this case the character curves are separated from each other very sharply and now the change of hydrophobicity coincides very well with the change of unsaturated character, but it has to be noted that the shape of this curves differs very much from the shape of the character curves of log k' values /see: Fig. 16a-d/. This is a definite sign that in the HPLC behaviour of these molecules, such steric and electronic forces may play a role which are not taken into account in the calculation of X^2 values. This is an establishment which seems to be particularly valid for the X^2 value of the derivatives with C_9-methyl group i.e. the last one caused an extremely high increase in log k' values.

REFERENCES

(1) MÉSZÁROS, Z., KNOLL, J., SZENTMIKLÓSI, P., HORVÁTH, G., HERMECZ, I: Arzneim. Forsch. 22 815 (1972)

(2) NÁRAY-SZABÓ, G., HERMECZ, I., MÉSZÁROS, Z.: J.C.S. Perkin 1853 (1974)

(3) PAPP, O., SZÁSZ, Gy., VALKÓ, K., HANKÓ-NOVÁK, K., HERMECZ, I. KÖKÖSI, J.: Acta Pharm. Fenn. 90 107 (1981)

(4) HANAI, C., TRAN, J., HUBERT, J.: J. of High. Res. Chromatography and Chrom. Comm. 4 454 (1981)

(5) KÖKÖSI, J., HERMECZ, I., SZÁSZ, Gy., MÉSZÁROS, Z. :Tetrahedron Letters 22 4861 (1981)

(6) HANKÓNÉ NOVÁK, K., SZÁSZ, Gy., PAPP, O., VÁMOS, J., HERMECZ, I.: Acta Pharm. Hung. 51 246 (1981)

ANALYSIS OF AMINO ACIDS

Proc. of the Symp. on Advances in Liquid Chromatogr., Szeged, Hungary, 1982

SCREENING FOR AMINO ACID METABOLISM DISORDERS BY ION-EXCHANGE THIN-LAYER CHROMATOGRAPHY

J. KOVÁCS,* P. KISS**

*Section of Pediatrics, Péterfy Hospital, Budapest
**Apáthy István Children's Hospital, Budapest, Hungary

In 1908 Sir Archibald Garrod the famous English physician introduced the concept of inborn errors of metabolism to the medical sciences. Since that time tremendous progress has been made in this field, mainly due to the development of effective analytical techniques.

Among the various forms of inborn errors of metabolism, the amino acidopathies are in the foreground, for several reasons:

1/ the amino acidopathies are the most frequent inborn errors, e.g. the incidence of phenylketonuria /PKU/ is generally 1:10000 in newborn infants.

2/ By developing modern analytical methods the detection of this type of disorders seems to be the easiest.

3/ In the case of an early diagnosis, the development of the disease - an otherwise incurable mental retardation - can be prevented by adequate treatment and the patient can be raised as a normal and healthy individual.

To illustrate the importance of this problem it should be mentioned that until 1978 more than 30 million newborn babies were tested throughout the world. Screening at this scale requires highly productive laboratory routine methods. Ion-exchange TLC of amino acids on Fixion sheets (1), owing to its simplicity and reliability, meets these demands. Between 1972 and 1980, more than 18 000 samples were analysed in the authors, laboratory with these techniques: the present paper reports on the main experiences of this work.

METHODS

The purpose of a screening program is to check all newborn babies in their first week of life, for only the biochemical alterations indicate an amino acid disorder.

Blood samples dried on filter paper are collected and sent to the laboratory (2,4). Discs are punched from the filter paper and placed into tubes. The elution of amino acids is performed with alcoholic hydrochloric acid solution and the eluate is transferred onto the chromatosheet. To simplify and to accelerate the procedure, small size 10 x 10 cm Fixion 50 x 8 sheets are used. With a pH = 5.2 citrate buffer the six most important amino acids, namely arginine, histidine, lysine, phenylalanine, tyrosine and leucine-izoleucin-valine can be well separated in about 60 minutes. In normal conditions the blood level of the various amino acids is similar and constant, therefore pathologic increases - such as with PKU - can easily be detected.

Practically the same procedure is applied for the detection of branched chain amino acids too (5). Using pH 3.3 citrate buffer, clear-cut separation of leucine-isoleucin and valine could be achieved.

Detection of amino acids in urine has also an important role in clinical practice. For this purpose desalting or deproteinization of the urine samples is not necessary. pH 3.3 citrate buffer as a solvent on the conventional chromatosheets ensures suitable separation of the urinary amino acids.

An important advantage of the Fixion method is that almost all the most important and most frequent disorders or amino acid metabolism can be detected and recognized /Table 1/. Among them obviously the PKU has the greatest importance, but in certain patients the early diagnosis of a life threatening metabolic disorder has considerable significance.

RESULTS AND DISCUSSION

Between 1972 and 1980 more than 18000 blood samples were analyzed. The results are summarized in Table 2.

To illustrate the practical consequences, the following interesting cases can be mentioned.

TABLE 1

Detectable disorders by Fixion method

Diagnosis	Amino acids in plasma	urine	Enzyme deficiency	Incidence	Clinical features
Phenylketonuria	PHE	o-PHE-acetic acid	PHE-OH-ase	1:10000	Mental retardation Convulsion Eczema
Hyperphenyl-alaninemia	PHE			1:40000	
Tyrosinemia /subtypes/	TYR	TYR	p-HPPA-oxidase amino transferase		Failure to thrive Richner-Hanhart sy. Transient forms
Hyperlysinemia	LYS $/NH_3/$	LYS N-ac-LYS	LYS-acylase LYS-dehydrogenase LYS-keto-glut-red.		Convulsions Coma Muscular hypotony Progression
Histidinemia	HIS	HIS ALA	HIS-ammonialyase	1:15000 /?/	Speech disorders Mental retardation Therapy /?/
Maple syrup Urine disease	LEU ILE VAL	LEU ILE VAL	Branched chain ketoacid decarboxylase	1:200000 - 1:500000	Types: 1. classical form 2. intermittent form 3. mild variant
Hyperleucinemia	ILE LEU	--	LEU, ILE-transaminase		Fatal outcome
Hypervalinemia	VAL	VAL	VAL-transaminase		Failure to thrive Neural disorders
Cystinuria	-	CYS			3 types Reanal calculi

Among the screened newborns one case of PKU was discovered on the 5th day of life. The chromatogram shows the striking spot of elevated blood phenylalanine level /Fig.1/. The diagnosis was confirmed at 2 weeks of age and phenylalanine free diet was introduced. The somatic and metal development of the baby turned to be completely normal.

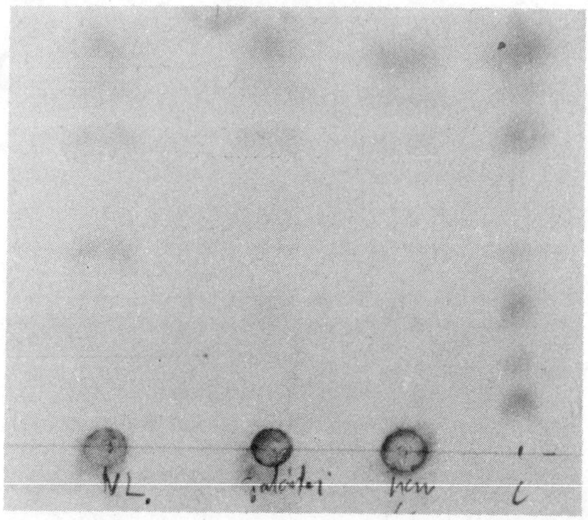

Figure 1. Detection of PKU on 10x10 cm Fixion 50x8 chromatosheet using citrate buffer pH 5.23 N.L.: patient with PKU

A case of leucinemia was discovered analysing the blood sample of another newborn baby /Fig.2/. Untreatable convulsions and metabolic acidosis were the main clinical symptoms. Although correct diagnosis was established, the therapeutic efforts failed and the fatal outcome could not be prevented. However, regarding the following pregnancies of the mother, prenatal diagnosis may be available.

Special interest is directed to the question of neonatal tyrosinemia. This disorder is mostly transient, but occasionally long time therapy is necessary. Therefore regular control of the blood tyrosin level is obligatory.

Blood samples are regularly sent to us from patients with mental retardation. In these cases establishing the correct diagnosis enhances the possibilities of suitable treatment.

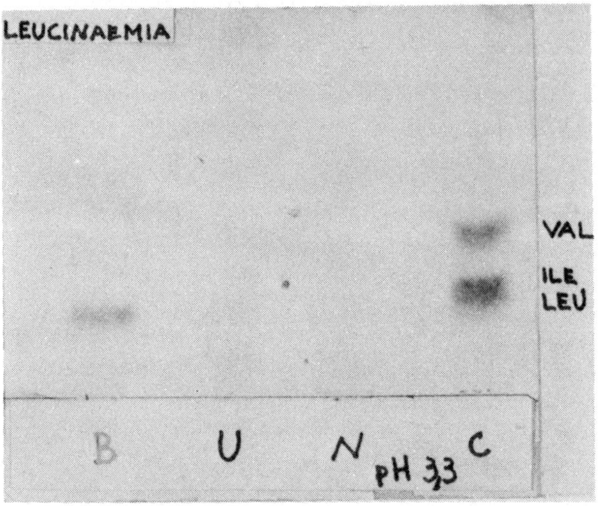

Figure 2. Detection of leucinemia using 10x10 cm Fixion 50x8 chromatosheet, citrate buffer pH 3.3. B: blood sample of leucinaemic newborn. U: urine of the same patient N: normal blood serum

A 14 year - old girl was observed throughout her life with somatic and mental retardation, skin disorders including scleroderma. The latter condition led to walking disability. From one drop of blood the whole clinical picture was clarified. The child had PKU and her skin disorder is a known consequence of this disease if untreated. Introducting low phenylalanine diet resulted in a striking improvement in her condition.

Three cases of maple syrup urine disease variants were detected. The first case (6) a three year-old mentally retarded boy was referred to us by the genetic counselling service. In his blood sample there was a moderate rise in branched chain amino acids. In order to establish the type of disorder a protein loading was performed. The results could be clearly demonstrated on the chromatogram /Fig.3/ as well as by quantitative analysis. The next figure displays the changing pattern of branched chain as well as basic amino acids, expressed in per cents of the fasting levels: curve 1: branched chain amino acids in patient, 2: in healthy protein load in a patient of maple syrup urine disease variant: distribution of branched chain amino acids in blood serum control, 3: basic amino acids in patient and 4: in control.

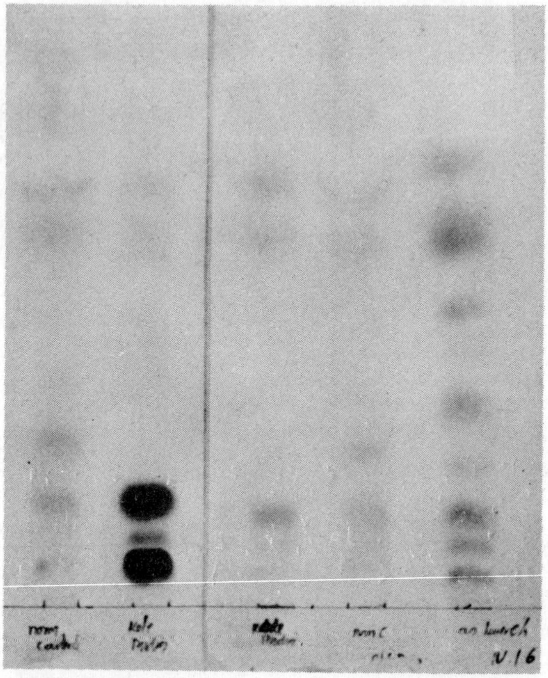

Figure 3.

It could be observed that the level of branched chain amino acids persisted in the pathological range, failing to return to the initial level, whereas no change occurred in the pattern of the other amino acids /Fig.4/.

Almost the same laboratory and clinical findings were found in two unrelated girls. The latter displayed cranio--facial dysmorphism, resembling the Weaver syndrome.

In the management of these patients low protein diet was introduced, complemented with leucin, isoleucin, valine free formula. Marked improvement was achieved in their mental and somatic development.

Recently much attention has been focussed on the congenital disturbances of the urea cycle. Five main enzymatic defects are responsible for almost similar clinical consequences.

TABLE 2

NEWBORN SCREENING		20 000
PKU	2	
Leucinemia	1	
Tyrosinemia	2	
Transient tyrosinemia	5	
MOTHERS OF MICROCEPHALIC CHILDREN		13
Hyperphenylalaninemia	2	
SCREENING OF MENTALLY RETARDED:		350
PKU-hyperphenylalaninemia	16	
MSUD variants	3	
Lysinemia	1	
Alaninemia /pyruvate--carboxylase def./	1	
Generalized amino acidemia	5	
SECONDARY DISTURBANCES OF AMINO ACID METABOLISM		8
Galactosemia	5	
Fructose intolerance /Tyr-Met-emia/	1	
CPS-deficiency /Arg-Orn-uria/	1	
D-resistant rickets	1	
Glycinuria	1	

Serious symptoms of the central nervous system arise already in early infancy. Their common biochemical background, the elevated blood ammonia level is responsible for the clinical picture.

A 3 month - old boy was admitted to the hospital in comatous state, which was preceded by acute febrile illness and vomiting. Hepatic failure was suspected which was accompanied by elevated blood ammonia level. In this life threatening condi-

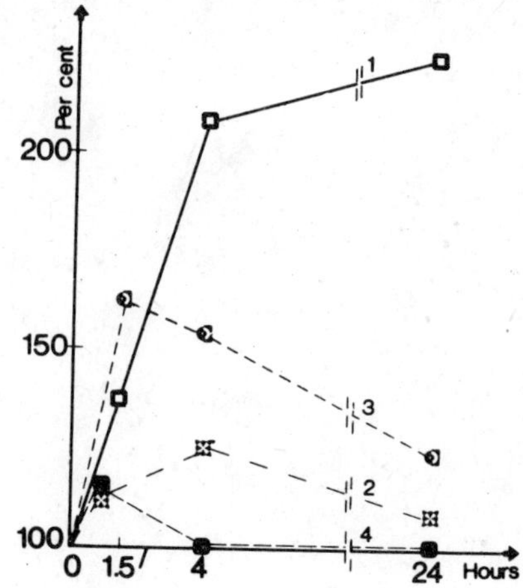

Figure 4. Serum amino acid pattern after 3g/kg protein load in MSUD and healthy control. The changing pattern of all branched chain as well as basic amino acids is expressed in per cents of the fasting levels. Curve 1. branched chain amino acids in MSUD; 2. branched chain amino acids in control; 3.basic amino acids in MSUD; 4. basic amino acids in control

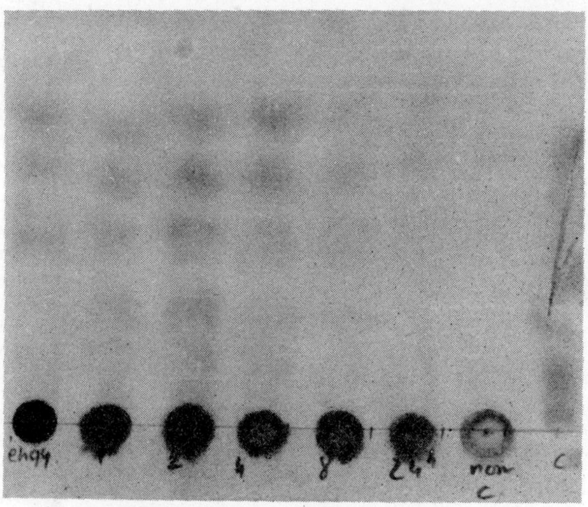

Figure 5. Increased urinary excretion of arginine and ornithine-lysine in a patient with carbamyl--phosphate synthetase deficiency

tion the rapid amino acid analysis of blood and urine led to the correct diagnosis. The blood amino acid level was normal, but in the urine, a marked dibasic amino aciduria, arginin- and ornithinuria were discovered /Fig.5/. This finding raised the possibility of the disturbance of the urea cycle, first of all carbamyl--phosphate synthetase deficiency. Protein restriction resulted in dramatic improvement in the patient's condition and the dibasic amino aciduria disappeared. Direct enzyme assay from liver biopsy specimen verified our supposition.

CONCLUSIONS

The advantages of the Fixion method in clinical practice may be summarized as follows:

1/ From a single drop of blood the complete amino acid pattern can be detected (1,2,4).

2/ The use of one dimensional, ascending technique with aqueous solvent systems makes the method simple, rapid, inexpensive and reproducible.

3/ Salts, urea, bilirubin or other substances which are present in biological material do not interfere.

4/ Deproteinization of the blood samples is necessary, but this could be achieved by very simple procedures.

5/ Under normal circumstances certain amino acids are present in almost identical concentrations in the blood, so that on the chromatogram the intensity of the histidine or tyrosine may serve as internal standards.

6/ The sensitivity of the method is satisfactory. For example, in the case of phenylalanine its concentration of 40 μmoles /liter can be easily detected.

7/ The method is suitable both for large-scale screening purposes, and for dietary control, since the blood samples may be sent by mail to the laboratory.

8/ In uncertain cases, the diagnosis must be confirmed by loading tests. The blood samples dried on filter paper can be collected and the analysis by means of the mentioned methods confirms or denies the diagnosis.

9/ The Fixion method is suitable for the analysis of urine and amniotic fluid (3).

10/ Finally, a quantitative evaluation is also possible by *in situ* densitometry (7).

ACKNOWLEDGEMENTS

Thanks are due to Dr. Tibor Dévényi for his help and interest and to Mrs. Ibolya Midrák for technical assistance.

REFERENCES

(1) DÉVÉNYI,T.: Thin-layer ion exchange chromatography on resin coated chromatoplates. I. Separation of aromatic and basic amino acids (phenylketonuria-test). Acta Biochim.Biophys.Acad.Sci.Hung. 5. 435 (1970).

(2) DÉVÉNYI,T., BÁTI,J., KOVÁCS,J., KISS,P.: Thin-layer ion-exchange chromatographic screening test for aminoacidemias in blood samples dried on filter paper. Acta Biochim.Biophys. Acad.Sci.Hung. 7. 237 (1972).

(3) DÉVÉNYI,T.: Thin-layer ion-exchange chromatography on resin coated chromatoplatesP a new tool in biochemistry, Hung.Sci.Instr. 30. 13 (1974).

(4) GUTHRIE,R., SUSIE,A.: A simple phenylalanine method for detection of phenylketonuria in large population of newborn infants. Pediatrics 32. 338 (1963).

(5) KOVÁCS,J.: Ion-exchange thin-layer chromatographic screening test for phenylketonuria and other aminoacidemias. Acta Pediatr. Acad.Sci. Hung. 14. 165 (1973).

(6) KOVÁCS,J.: Detection of maple syrup urine disease on resin coated chromato-sheets. Acta Biochim.Biophys. Acad.Sci.Hung. 14. 119 (1979).

(7) PONGOR,S., KOVÁCS,J., KISS,P., DÉVÉNYI,T.: Quantitative evaluation of thin-layer chromatograms by videodensitometry. II. Screening test for aminoacidemias in blood samples dried on filter paper. Acta Biochim.Biophys. Acad.Sci.Hung. 13. 117 (1978).

Proc. of the Symp. on Advances in Liquid Chromatogr., Szeged, Hungary, 1982

CORRELATION OF PHENYLALANINE AND TYROSINE VALUES DETERMINED BY CONTINUOUS FLOW FLUOROMETRY AND AMINO ACID ANALYZER

A. MRSKOŠ, R. POSPIŠIL, V. KOLCOVÁ

Pediatric Research Institute,
Brno, Czechoslovakia

The diagnosis of hyperphenylalaninemias and tyrosinemias,proper handling of phenylketonuric dietary therapy and phenylketonuric heterozygote determination is more accurate than phenylalanine and tyrosine determinations are. But high precision requires laborious and wearisome operations and provides few determinations only. For amino acids the appreciated reference method is column ionex chromatography namely with shortened programs. For the assay of many samples it is necessary to choose relatively precise mechanized methodologies e.g. continous flow fluorometry instead of manual operations.

The correlation was achieved on 20 serum samples with different levels of phenylalanine and tyrosine. Their determinations by ionex chromatography were performed by our own shortened program on the automatic amino acid analyzer AAA 881 (Mikrotechna,Prague) equipped with a 27 x 0.8 cm column packed with spherical resin Ostion LG KS 0802 (Lachema,Brno) to filling height of 25 cm. The flow rate of 0.35 M sodium citrate,pH 6.37, was 70 $ml.h^{-1}$ and that of ninhydrin 35 $ml.h^{-1}$. Column jacket temperature was 53 ^{o}C. We used our own modification of automated phenylalanine estimation by Hill et al. for continuous flow fluorometry and serum blanks determinations according to Ambrose's proposition for nonspecific fluorescence (Mrskoš,1976). We performed tyrosine estimation by a slight modification of Blau and Edwards's method (1971). For determinations we used AutoAnalyzer Mark I with moduls Sampler II and Fluometer II.

Methods elaborated in this way entitled us to carry out the correlation analysis to ascertain their accuracy,precision and mutual substitution. The equation $y = 0.96x + 0.52$ shows that for two methodologies followed, (y = ionex chromatography values, x = flow fluorometry values) there is linear dependency and the regression line is a straight one. The correlation coefficient $r = 0.9808$ is highly significant. Güttler and Hansen (1977) found by comparing manual fluorometric method and ionex chromatography the correlation coefficient 0.93 and regression line equation $y = 0.94x + 0.38$. Ambrose (1973) reported that his automated method for phenylalanine gave at the lowest phenylalanine level 6mg/dl by ionex chromatography, a significantly higher result 2.41 mg/dl. For higher phenylalanine levels the average values of his method were not significantly different from column chromatography. Although the correlation analysis is performed as a rule on larger sets and it is necessary to count in this case on a relatively broad reliability interval, it could be judged from the results that both methodologies for phenylalanine determination are in close direct dependence.

The second view on the values of both methodologies should answer the question if one methodology does not produce higher values than the other. By means of variance analysis, comparison of average values and confrontation with critical table values we came to the values stated below:

	x mg/dl	y mg/dl
Range	1.55 - 15.15	1.98 - 15.94
mean	6.37	6.63
SD^2	14.714	14.218

$$F = 1.035 \ (p > 0.05)$$
$$t = 0.212 \ (p > 0.05)$$

A probability higher than 0.05 ascertained by these tests informs us that the differences were not statistically significant. It means at the same time that both methodologies, column ionex chromatography and continuous flow fluorometry, are mutually replaceable and comparable for the determination of phenylalanine. Continuous flow fluorometry is then more efficient, speedy and cheaper, and as to the quantity of required blood, more considerate to the patient.

What has been already mentioned could also be appliėd for tyrosine determination:

regression line equation	$y = 0.83x + 0.42$
correlation coefficient	$r = 0.9968$

Güttler and Hansen found a correlation coefficient of 0.88 and their regression line equation was $y = 0.80x + 2.46$. The variance analysis was:

	x mg/dl	y mg/dl
Range	1.05 - 12.85	0.91 - 11.70
mean	5.52	5.00
SD^2	23.262	21.278

$$F = 1.093 \ (p > 0.05)$$
$$t = 0.336 \ (p > 0.05)$$

Tyrosine methodologies are mutually replaceable and comparable.

REFERENCES

AMBROSE,J.A. (1973) Report on a cooperative study of various fluorometric procedures and the Guthrie bacterial inhibition assay in the determination of hyperphenylalaninemia.Health Lab.Sci. 10 (3), 180-187

BLAU,K.,EDWARDS,D.J. (1971) A modified automated fluorometric method for the determination of tyrosine in blood and tissues, and its specificity. Biochem.Med. 5 (4),333-341

GÜTTLER,F., HANSEN,G. (1977) Serum tyrosine within the first hour after an oral load of phenylalanine. Scand.J.Clin. Lab.Invest. 37, 717-722

MRSKOŠ,A. (1976) A modified fluorometric method for blood phenylalanine determination on AutoAnalyzer I.Abstracts, 2nd Eur.Congr.Clin.Chem., Prague,Oct.3-8.1976.

Proc. of the Symp. on Advances in Liquid Chromatogr., Szeged, Hungary, 1982

ORAL LOADING TEST WITH L-METHIONINE

R. POSPIŠIL, A. MRSKOŠ, O. PODHRADSKÁ, O. ŠTOURACOVÁ

Pediatric Research Institue,
Brno, Czechoslovakia

Homocystinuria comprises a group of autosomal recessive disorders of sulphur amino acid metabolism. The most common type is due to a deficiency of cystathionine synthase, a pyridoxal phosphate-dependent enzyme which catalyses the reaction of serine with homocysteine to form cystathionine. This enzyme defect results in raised blood and urine levels of homocystine, methionine, cysteine-homocysteine disulphide and other sulphur amino acids and in lack of brain cystathionine.

It has been shown that heterozygotes for homocystinuria have approximately half the normal cystathionine synthase activity in liver biopsy specimens. This reduced ability of heterozygotes to metabolize methionine has been tested by a number of investigators in order to devise a test for the carrier state in this condition.

Dunn, Perry and Dolman measured plasma methionine concentrations in 4 normal controls and 4 parents after an L-methionine load following a 14-hour fast. Though there was no overlap of the range of methionine concentration in the two groups at 4 hours post load, the difference between ranges was insufficient to make this a reliable indication of the carrier state. Chase, Goodman and O'Brien determined plasma methionine in 4 parents and 8 normal subjects. Though a distinction between the two groups was noted, the number of heterozygotes studied was small and the value of this parameter for detection of heterozygotes is doubtful.

In the present paper the authors describe their results obtained from the evaluation of the tolerance test by means of methionine load in a patient with homocystinuria, in her parents and in a group of healthy children.

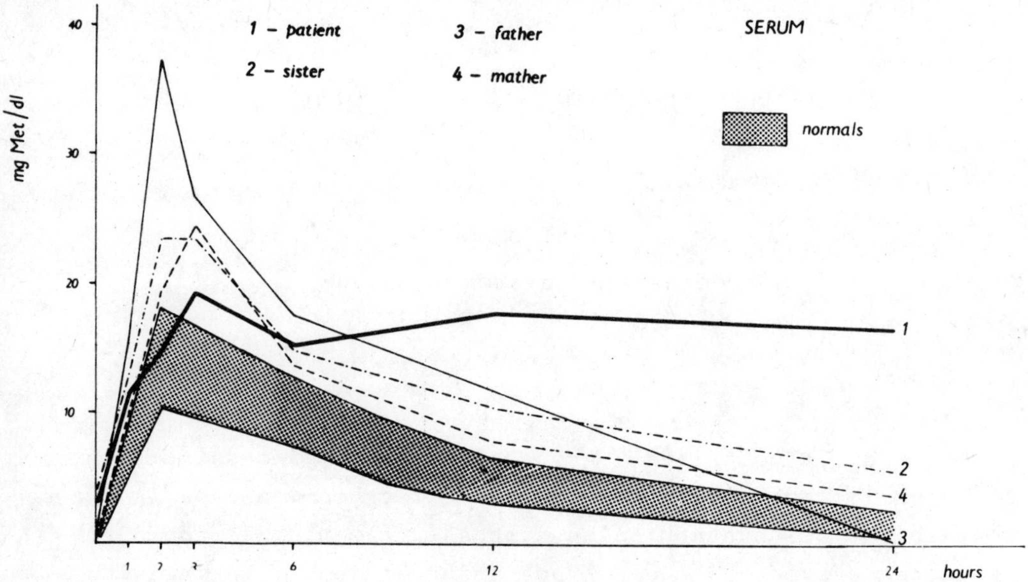

Figure 1

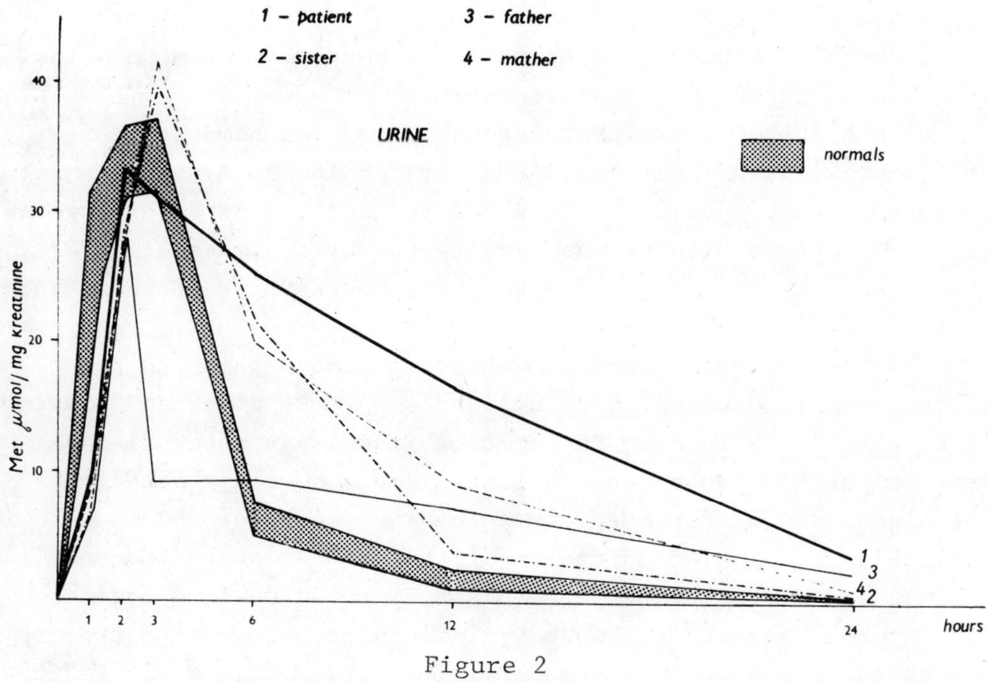

Figure 2

We administered 100 mg L-methionine per kilogram of body weight for an oral loading test. Urine was collected 24 hours before the challenge and venous blood samples were taken at 0, 1, 2, 3, 6, 12 and 24 hours after methionine administration.

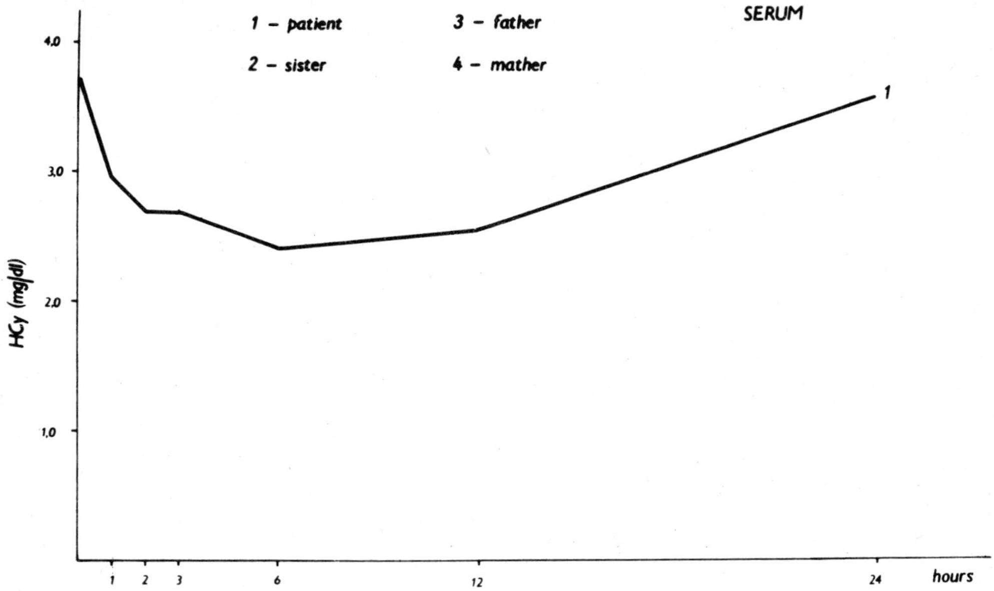

Figure 3

Methionine, homocystine and their oxidation products, further disulphide cysteine-homocysteine were estimated in serum and urine using a short programme on an amino acid analyzer.

In Fig.1. the menthionine values can be seen in serum of the patient her parents and sister and in a group of healthy children during the tolerance tests. In normal patients the maximum concentration of methionine was observed at the second hour, then the curve decreased until 24 to 48 hours when it reached the starting level. The patient did not return to the fasting value and her maximum was not too different from the normals. Her parents and sister had quite distinct increases and maximum and slower decrease than normals. The starting methionine value was reached in the father only. According to the literature it was possible to discriminate heterozygotes by higher fasting methionine levels. This was also the case in our study. It was interesting that the fasting value in proband was lower than in her mother. Of course her fasting homocystine level was high.

In Fig.2. the course of methionine excretion in urine of heterozygotes and proband can be seen after the loading test marked by a slower decline from the maximum against the normal values.

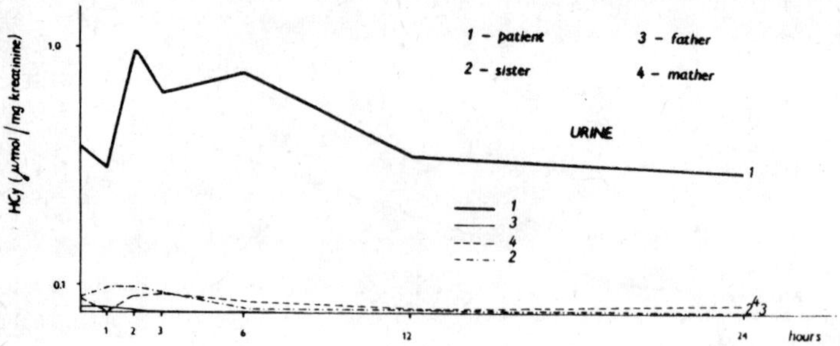

Figure 4

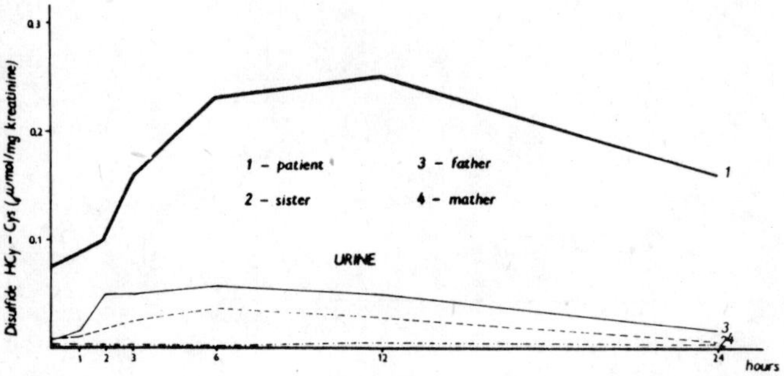

Figure 5

In Fig.3. serum homocystine was found in the patient only - that is before and after the load. In the urine, homocystine (Fig.4) was found in a higher concentration also in the patient only and in parents and sibling traces of homocystine were found. The most relevant result according to our experience was observed in the behaviour of mixed disulphide cysteine-homocysteine in urine before and after the methionine administration.

This disulphide (Fig.5) in urine of homocystinurics rose slowly to the twelfth hour and reached the starting concentration after some days. A similar course but in lower concentraions was found in heterozygotes also.

REFERENCE

(1) DUNN,H.G.,PERRY,T.L.,DOLMAN,C.L.: Homocystinuria a recently discovered cause of mental defect and cerebrovascular thrombosis. Neurology 16 407, /1966/

(2) CHASE,H.P.,GOODMAN,S.I.,O'BRIEN,D.: Treatment of homocystinuria. Arch.Dis.Child. 42 514 /1967/

(3) SARDHARWALLA,I.B.,FOWLER,B.,ROBINS,A.J.,KOMROWER,G.M. Detection of heterozygotes for homocystinuria. Study of sulphur-containing amino acids in plasma and urine after L-methionine loading. Arch.Dis.Child. 49 553 /1974/

Proc. of the Symp. on Advances in Liquid Chromatogr., Szeged, Hungary, 1982

THE METABOLIC CHANGES OF AMINO ACIDS IN MATERNAL BLOOD AND MILK DURING PREGNANCY OF HEALTHY AND PHENYLKETONURIC MOTHERS

J. HYÁNEK, H. VILETOVÁ, V. TRNKA, V. KUNOVÁ, J. ČERVENKA

Medical Faculty, Charles University, Prague, Czechoslovakia

INTRODUCTION

Pathologically increased concentration of phenylalanine in maternal blood has a detrimental effect on the early development of the fetus. Its manifestations are low birth weight,microcephaly, congenital heart disease, multiple bone deformities, mental retardation etc. (Waisman 1967,Hansen 1970, Hyánek 1979).

MATERIAL AND METHODS

Changes of amino acids have been studied in maternal blood, amniotic fluid, umbilical blood and maternal milk of 6 mothers suffering from different forms of phenylketonuria and in 6 healthy pregnant or breast-feeding mothers.

Samples of venous blood were deproteinized with the crude sulphosalicylic acid (300 mg/ml of serum), centrifuged and 200-400 µl of supernatant were applied to the column of amino acid analyzer Mikrotechna AAA 881. Long programme for the whole spectrum as well as the short programme for the aromatic amino acids were used (Spackman 1958, Benson 1969). Samples of maternal milk were delipidized and deproteinized before analysis on the same analyzer. 2 ml of milk were twice treated with 2 ml of chloroform, centrifuged at 2500 g for 10 min and the supernatant was deproteinized, 400 µl of supernatant was applied on the column.

In 3 healthy breast-feeding mothers the modified peroral loading test with L-phenylalanine after Guttler and Hansen was performed (Hyánek 1981). Blood and milk samples were taken before the loading and at 1,2,3 and 4 h after peroral loading with L-β-phenylalanine (Merck FRG) 0.6 mmol per kg of body weight. Direct measurement of the phenylalanine-hydroxylase activity in tissue from a liver biopsy was not performed.

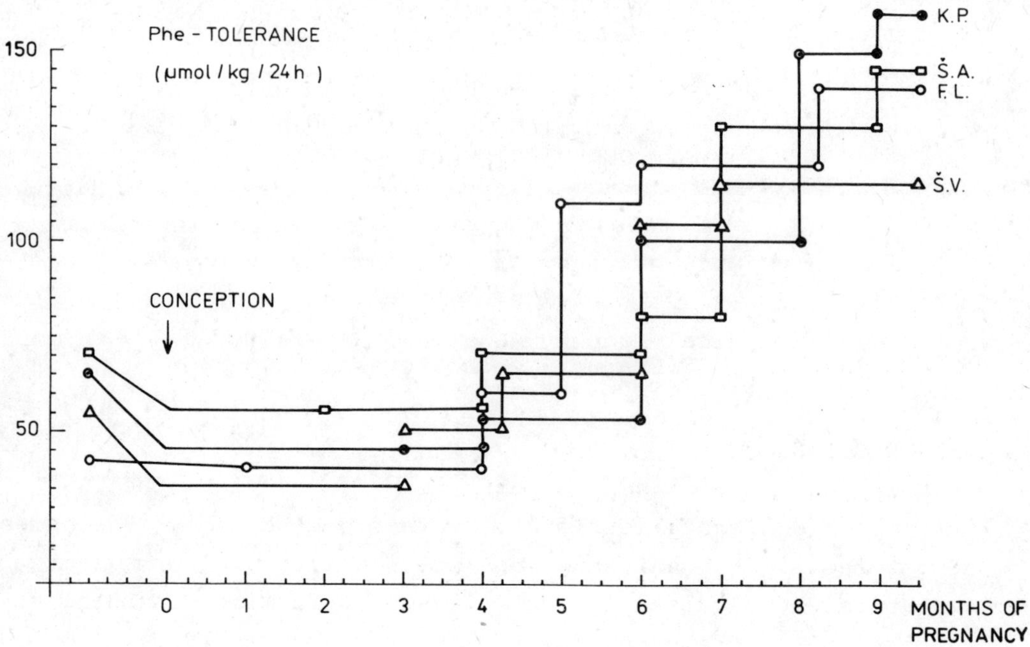

Figure 1. The change of phenylalanine tolerance during pregnancy of phenylketonuric mothers

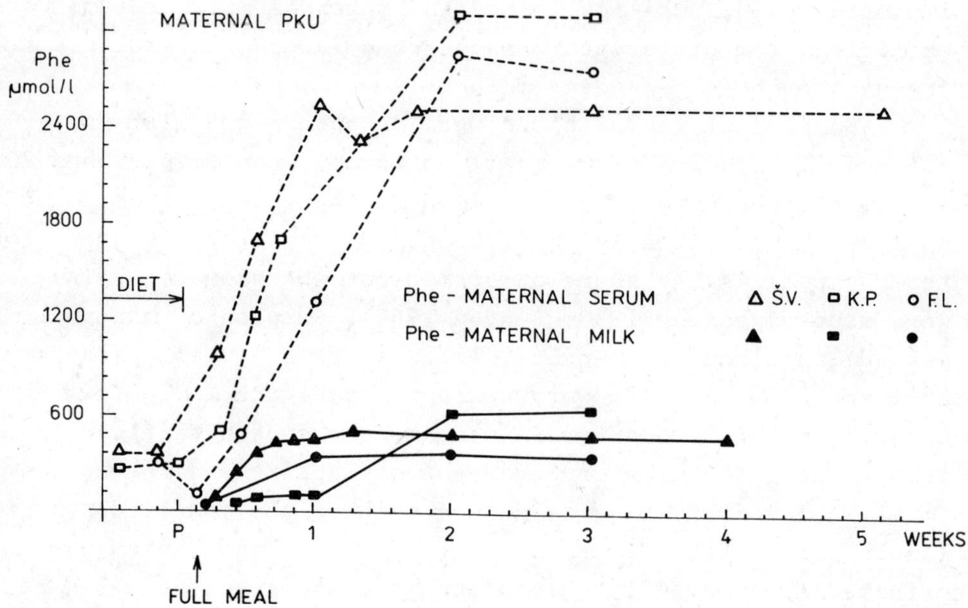

Figure 2. The phenylalanine levels in maternal blood and maternal milk after delivery of phenylketonuric breast-feeding mothers.

RESULTS

A healthy placenta increases the amino acid level in the fetal blood due to its sucking effect. The feto-placental quotient (FPQ) for the most important amino acids ranges between 0.5 to 4.8, with an average value of 2,2. The highest values were recorded for glycine, taurine, arginine while the lowest values were recorded for alanine, leucine, histidine. FPQ of aromatic amino acids (tyrosine + phenylalanine) in our groups of investigated pregnant mothers are shown in Table 1. Our findings are in agreement with those reported earlier by Lindblad (1967) and Butterfield (1963). At the end of pregnancy many FPQ show a decreasing tendency in correlation to the decreasing function of placenta (Hyånek 1982).

In correlation to the decreasing values of phenylalanine FPQ in phenylketonuric mothers, the phenylalanine tolerance increases in the administered food*. Its value increases 3-4 times from the second half of pregnancy as shown in Fig.1.

The amino acid levels in maternal milk of phenylketonuric mothers after delivery do not correlate to the values in maternal blood. They hardly attain 1/4 or 1/5 of the mother's phenylalanine blood level as shown in Fig.2. The mammary secretory mechanism probably yields some protective effects on the phenylalanine secretion. The same discrepancies of amino acids between the maternal blood and maternal milk were observed also in the group of our healthy breast-feeding mothers after L-phenylalanine load (Fig.3). The milk-serum quotient reaches the lowest value for phenylalanine /0.2/ and tyrosine /0.3/. While for the other amino acids its value ranges from 0.5 to 0.7.

---

* Phenylalanine tolerance means the amount of phenylalanine in µmol/kg body weight/day that can be metabolized without increasing the therapeutic level of phenylalanine in maternal blood (i.e. over 400 µmol/l).

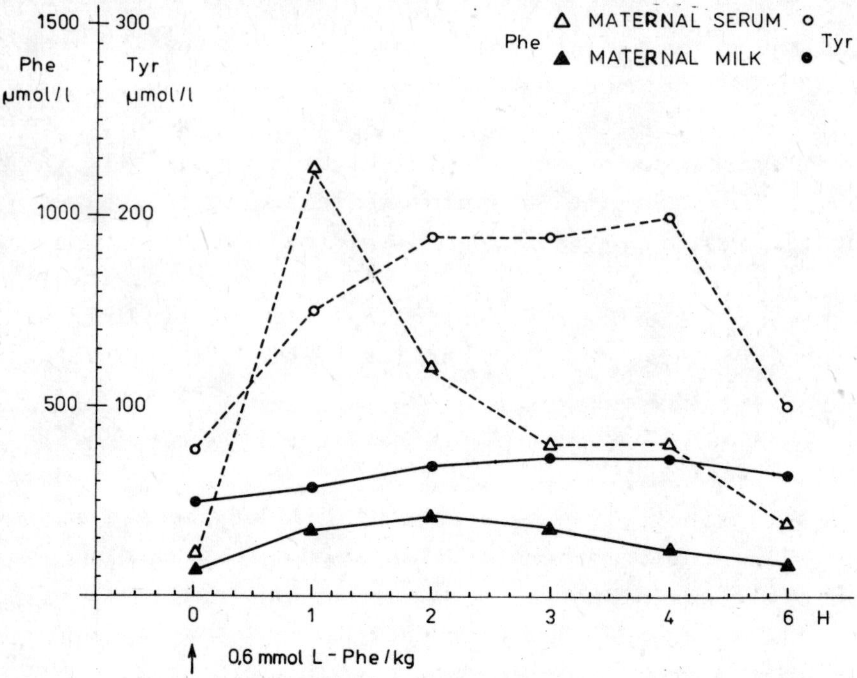

Figure 3. The levels of phenylalanine in maternal blood and maternal milk after L-phenylalanine loading in healthy breast-feeding mothers

TABLE I

Fetoplacental quotient /FPQ/ of aromatic amino acids in pregnant women suffering from phenylalanine disturbancies /36-40th week of pregnancy/

Patient	FPQ of Phenylalanine	FPQ of Tyrosine
I. Phenylketonuria:		
Š.Z.	2,62	1,42
K.P.	1,50	1,00
Š.H.	1,60	1,40
R.M.	1,60	1,70
Š.V.	2,0	1,40
II. Hyperphenylalaninemia		
V.H.	1,18	3,25
M.E.	3,40	-
K.H.	2,55	0,89
K.L.	2,61	2,90
III. 6 Healthy Controls	2,10 ± 0,5	1,72 ± 0,6

DISCUSSION

The finding of different levels of phenylalanine in maternal blood and maternal milk is of great importance for the breast-feeding of children born from phenylketonuric mothers. Some authors /Fisch 1967, Pueschel 1977/ do not recommend breast-feeding with own maternal milk as a possible reason for increase of phenylalanine in the newborn's blood. Using L-phenylalanine tests, in controlled genetic prevention, we can find healthy partners for intended marriage to phenylketonuria treated women. According to the valid genetic laws, such a marriage of a woman homozygote having phenylketonuria with a man who is a healthy homozygote, only produces a heterozygote child with sufficient enzyme system of phenylalanine-hydroxylase. The phenylalanine tolerance of the pregnant woman has to be improved in the second half of pregnancy. Watching the increasing phenylalanine tolerance during pregnancy we have no doubts, that a phenylketonuria heterozygote will be born, and we therefore allow the mothers to breast-feed their baby and thus to enjoy the happiness of motherhood as the small reward for the drastic dietetic therapy during the whole pregnancy.

SUMMARY

Free amino acid in maternal blood, fetal blood, amniotic fluid and maternal milk have been studied by means of column chromatography on amino acid analyzer Mikrotechna AAA 881. The amino acid levels in maternal milk do not attain the values of maternal or fetal blood in the group of breast feeding mothers suffering from phenylketonuria. The same findings were observed in healthy breast-feeding mothers after L-phenylalanine load.

REFERENCES

BENSON,J.V.,GORDON,M.J.,PETTERSON,J.A.: Accelerated chromatographic analysis of amino acids in physiological fluids. Anal.Biochem. 18 228-240 (1967)

BUTTERFIELD,L.J., O'BRIEN,D.: The effect of maternal toxaemia and diabetes on transplacental gradients of free amino acids. Arch.Dis.Child. 38 326-327 (1963)

FISCH,R.Q., JENNES,S.: The effect of excess L-phenylalanine on mothers and on their breast fed infants. J. Pediat. 71 176-180 (1967)

HANSEN,H.: Epidemological considerations on maternal hyperphenylalaninemia. Amer.J.Ment.Defic. 75 22-26 (1970)

HYÁNEK,J., HOMOLKA,J.,TRNKA,V.: Results of screening for phenylalanine and other amino acid disturbances among pregnant women. J. Inher.Metab.Dis. 2 59-63 (1979)

HYÁNEK,J., VILETOVÁ,H., KUNOVÁ,V.: Phenylalanine loading test in genetic conselling: 5 years' experience with its premarital use. J. Inher.Metab.Dis. 4 51-52 (1981)

HYÁNEK,J.: Some metabolic studies on amino acids during pregnancy. Proc. of the 20th Ann.Symp. of SSIEM, Manchester, 1982

LINDBLAD,B.S., BALDESTEN,A.: The normal venous plasma-free amino acid levels of non-pregnant women and of mother and child during delivery. Acta Paediat.Scand. 56 37-48 (1967)

PUESCHEL,S.M., HURN,C., ANDREWS,M.: Nutritional management of the female with phenylketonuria during pregnancy. Amer.J. Clin.Nutr. 30 1153-1160 (1971)

SPACKNAM,D.H.,STEIN,W.H.,MOORE,S.: Automatic recording apparatus for use in the chromatography of amino acids. Anal.Chem. 30 1190-1206 (1958)

WAISMAN,H.A.: Role of hyperphenylalaninemia in pregnant women as a cause of mental retardation in offspring. Amer.J. Obstet. Gynecol. 99 431-433 (1967)

ANALYTICAL AND PREPARATIVE SEPARATION OF PEPTIDES AND PROTEINS

Proc. of the Symp. on Advances in Liquid Chromatogr., Szeged, Hungary, 1982

LIQUID CHROMATOGRAPHY, THIN-LAYER CHROMATOGRAPHY AND HIGH-PERFORMANCE LIQUID CHROMATOGRAPHY OF OXYTOCIN, VASOPRESSIN, SOME OF THEIR SPECIFIC ANALOGUES AND FRAGMENTS

L. BALÁSPIRI,* M.V. TÓTH,* T. FEKETE,* T. JANÁKY,*
F.A. LÁSZLÓ,* G. TÓTH,** F. SIROKMÁN**

*Institute of Medicinal Chemistry and Endocrinology Unit, First Department of Medicine, Medical School, Szeged, Hungary
**Biological Research Center, Hungarian Academy of Sciences, Szeged, Hungary

INTRODUCTION

Oxytocine (OT) was the first biologically active peptide synthetized in the laboratory of du Vigneaud, in 1953. The vasopressins (VPs), 8-arginine vasopressin (AVP) and 8-lysine-vasopressin (LVP) are also nonapeptides which differ from oxytocine only in positions 3 and 8 with phenylalanine and arginine or lysine which replaced the isoleucine and leucine, respectively (Fig. 1).

These kinds of peptides, their analogues and fragments can be synthetized by two, essentially different routes: either by the conventional ways (1) or by the so called Merrifield solid phase peptide synthesis (2). All syntheses - except the preparation of the smallest fragments - are followed by very important purification steps using the classical liquid chromatography (LC) on different kinds of columns, thin-layer chromatography (TLC) and recently high-performance liquid chromatography (HPLC).

In our research group we began to work on this area in 1973. The first successful results were the synthesis of OT, AVP and six specific antidiuretic analogues of the latter compound (Fig. 2) on solid phase (3-6).

Since then we have synthetized these peptide hormones and the mentioned analogues as precursors of iodinated and tritiated derivatives (7-9) for metabolic studies (7-9). Several other selective AVP analogues have also been prepared (10-11).

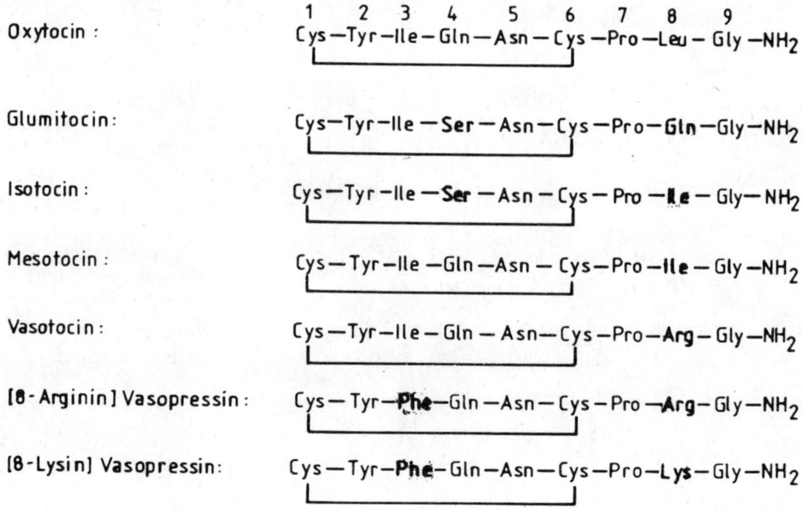

Figure 1 Structure of naturally occurring posterior pituitary peptide hormones isolated from its extracts

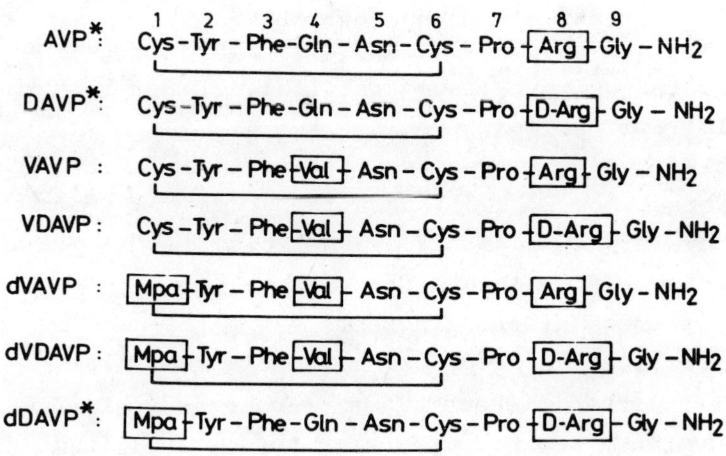

Figure 2 The structure of some specific antidiuretic (AD) AVP analogues. Mpa = mercaptopropionic acid, DAVP = (8-D-arginine)-vasopressin, dVDAP 1 (1-deamino, 4-valine, 8-D-arginine)--vasopressin, dDAVP = (1-deamine, 8-D-arginine)-vasopressin

All syntheses were carried out using the Merrifield solid phase peptide synthesis method modified by MANNING (12) and MEIENHOFER (13) for production of posterior pituitary hormones.

DISCUSSION

Three main consecutive steps are frequently used in the methods of preparation of peptides:

I. Synthesis of the protected amide,

II. Deprotection and oxidative formation of disulphide bridge,

III. Purification of the crude peptide amide by LC on Sephadex G columns and analytical control of the homogeneity of the pure peptide by TLC or HPLC.

Example for purification of OT and AVP analogues is gel filtration on Sephadex G-15 gel columns in two consecutive runs using acetic acid containing eluents (50% and 0.05 M aqueous solutions in the first and second steps of purification, respectively), the chromatogram can be seen in Fig. 4.

We have routinly used TLC as the control of purity of OT, LVP and AVP as well as their analogues. Table 1 summarizes the results from our earlier publications (3-6).

TLC data of several recently synthetized protected AVP, LVP and OVP (8-ornithine vasopressin) analogues (14) can be found in Table 2.

TEPLAN and his co-workers synthetized the tritiated LVP (^{3}H-LVP) for metabolic studies (15) which has been performed (16). After several weeks the originally pure product was partially destroyed because of known reasons. We successfully recovered the ^{3}H-LVP in full specific biological- and radioactivity by the help of a two-step micro procedure using carboxymethyl cellulose and CG-50 column chromatography (17). The elution curves with detection of biological- and radio-activity are shown in Figs 5, 6 and 7.

Up to now, tritiation of vasopressins and their analogues has not been perfectly solved (9-11). Recently we

Table 1

Analogues	System (R_f)									
	BAW(4:1:1)		BAW(4:1:5)		BAW (8:5:4)		BAW(30::20:6:24)		$CHCl_3$/MeOH (7:3)	
	prot.	free	prot.	free	prot	free	prot.	free	prot.	free
OT	-			0,47					0,6	
AVP	0,50			0,40		0,19		0,55	0,63	
LVP	0,50			0,40						
DAVP	0,55					0,17		0,47	0,63	
dDAVP	0,60					0,35		0,53	0,67	
VAVP	0,60	0,20				0,20		0,60	0,48	
VDAVP		0,17							0,96	
dVAVP	0,60	0,21				0,40		0,61	0,45	
dVDAVP	0,65	0,22				0,45			0,70	

Table 2

Analogues	System (R_f)	
	BAW (4:1:1)	BPAW (30:20:6:24)
Pip^7AVP	0,81	0,65
$dPip^7AVP$	0,68	0,62
$D\text{-}Pip^7AVP$	0,65	0,60
$dDPip^7AVP$	0,63	0,60
$/desGly\text{-}NH_2/^9dVDAVP$	0,71	0,72
$/desGly\text{-}NH_2/^9dDAVP$	0,70	0,65
$/desGly\text{-}NH_2/^9AVP$	0,67	0,71
$/desGly\text{-}NH_2/^9LVP$	0,67	0,71
$/desGly\text{-}NH_2/^9OVP$	0,47	0,72

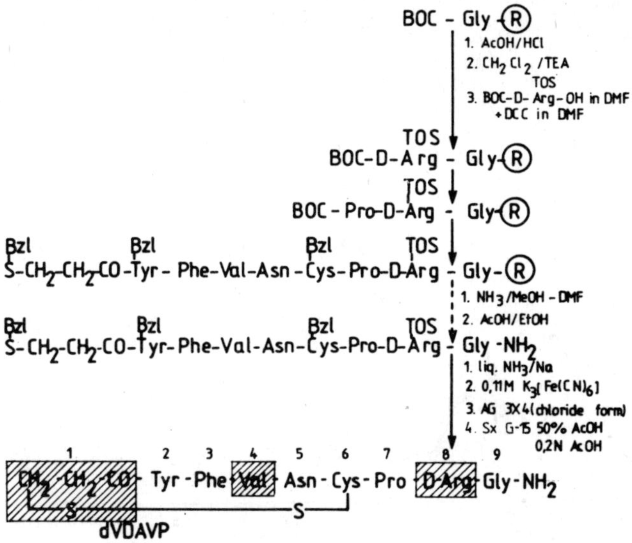

Figure 3 Synthesis of (1-deamino, 4-valine, 8-D-arginene)-vasopressin (dVDAVP) by Merrifield peptide synthesis on solid resin

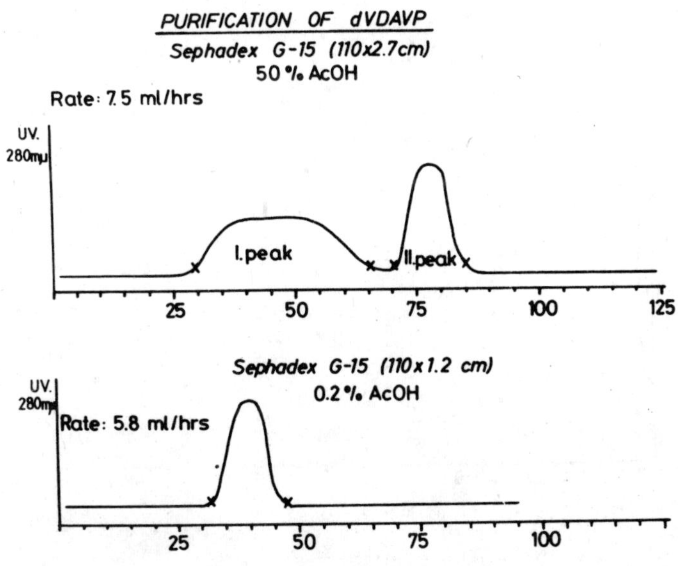

Figure 4 Purification of crude (1-deamino, 4-valine, 8-D-arginine)-vasopressin by gel filtration. Peaks were detected at 280 nm, flow rate was 7.5 ml/h and 5.8 ml/h in the experiments using the 50% and 0.2% acetic acid, respectively

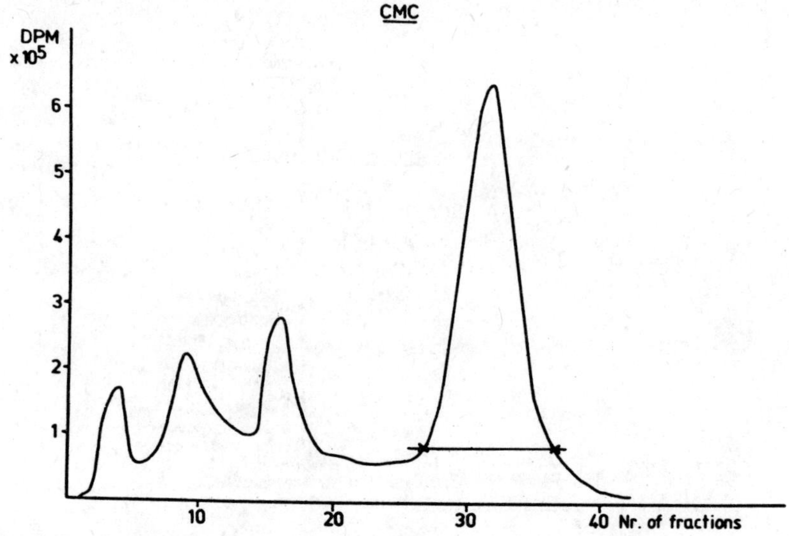

Figure 5 Elution curve of crude ^{3}H-LVP from carboxymethyl cellulose column

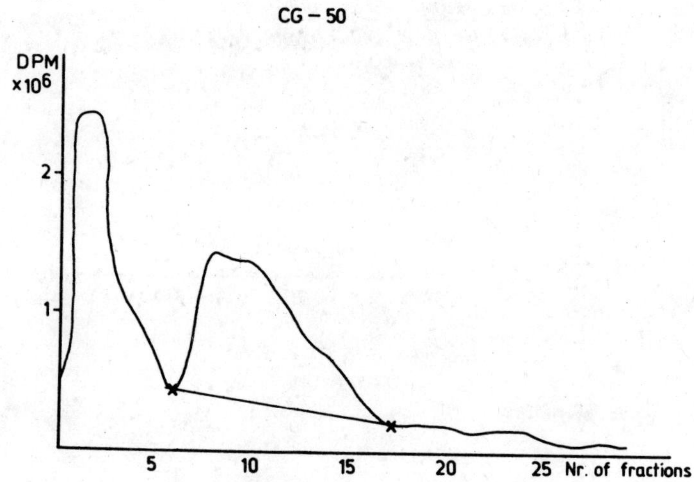

Figure 6 Elution curve of crude ^{3}H-LVP from CG-50 column

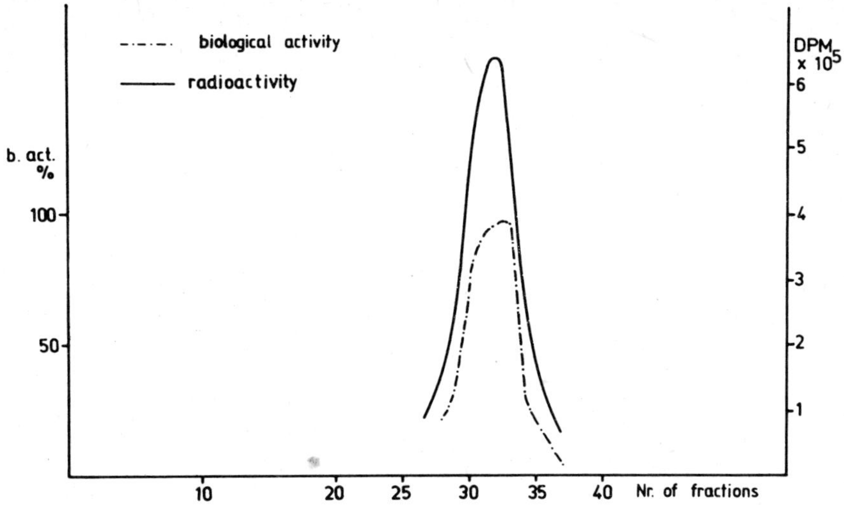

Figure 7 Purity control of recovered ^{3}H-LVP by measurement of radioactivity and ADH-activity

applied different approaches for two-step tritiation of OT and AVP analogues starting from pure analogues using iodinated derivatives as precursors (7-11) which are different from any hitherto known synthetic way of ^{3}H-LVP (15). Without going into details an elution curve will be shown, obtained at the purification of iodinated LVP (18) on HPLC column (Fig. 8).

Separation of iodinated LVP and AVP analogues on Sephadex G columns (19), control on TLC plates and verification of the successful separation by LOWRY test and UV spectra will soon be published (9).

Recently the synthetic work of analogues of OT and VP has been turned to look for new, rather specific antagonists. Especially interesting results have emerged from Manning's laboratory, where the study of posterior pituitary hormones, their analogues, metabolism and receptors are in the limelight.

Recently, there has been an increasing interest in the purification and analytical control of medium sized peptides (between 300 and 3000 MW) performing the separation by reversed-

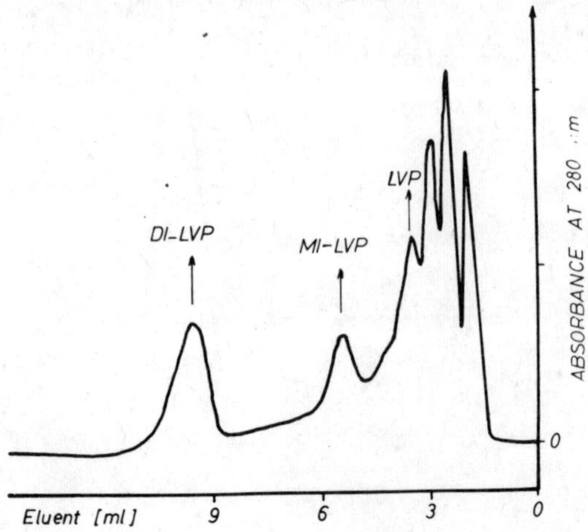

Figure 8 Elution curve of the separation of LVP, monoiodine-lysine-vasopressin (MI-LVP) and diiodine--lysine-vasopressin (DI-LVP) on u-Bondapack C_{18} column by HPLC (9)

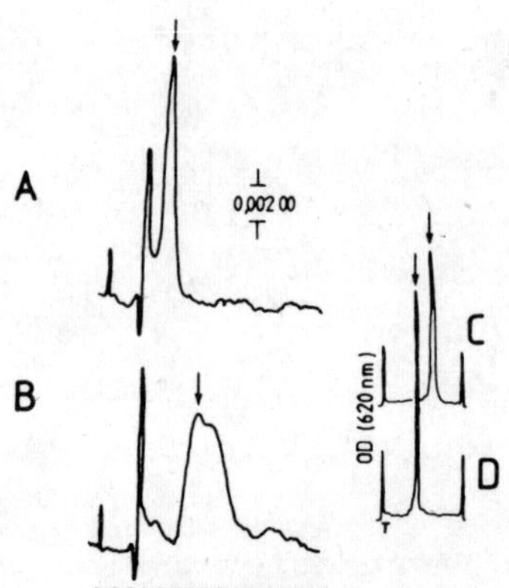

Figure 9 HPLC separations of Neurophysin I (a) and Neurophysin II.(B). Elutions were made by 51.6% methanol and 48.4% 0.01 M acetate buffer (pH 5.7) with 1 ml/min flow rate. Inset figures are their PAGE investigation (C and D) in 7.5% acrylamide gels using tris-glycine buffer at pH 7.9 (23)

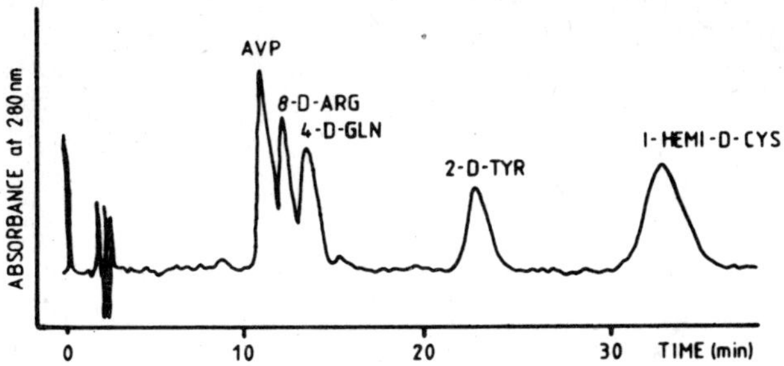

Figure 10 HPLC chromatogram of the AVP analogues on μ-Bondapack C_{18} column by elution of 0.075 M triethylamine - acetic acid buffer with 6% THF as organic modifier (25)

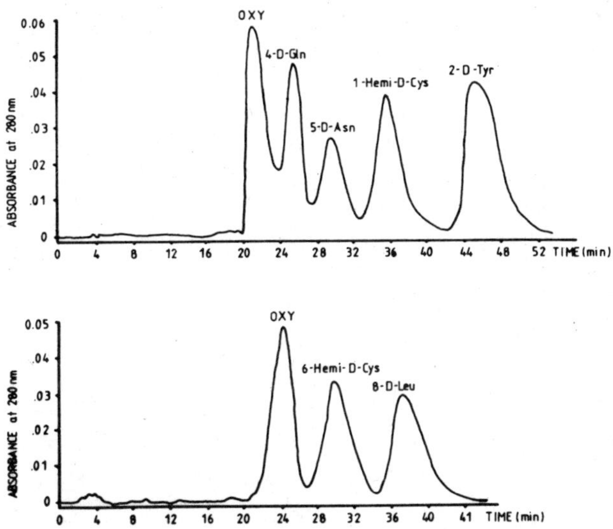

Figure 11 HPLC separation of OT and its D-analogues on μ-Bondapack C_{18} column. Elutions were made by 0,05 M ammonium acetate (pH 4) - tetrahydro furane 9:1 with flow rate = 1.6 ml/min an 0.01 M ammonium acetate (pH 4) - methanol 82:18 with flow rate 2 ml/min on the upper and bottom figures, respectively (24)

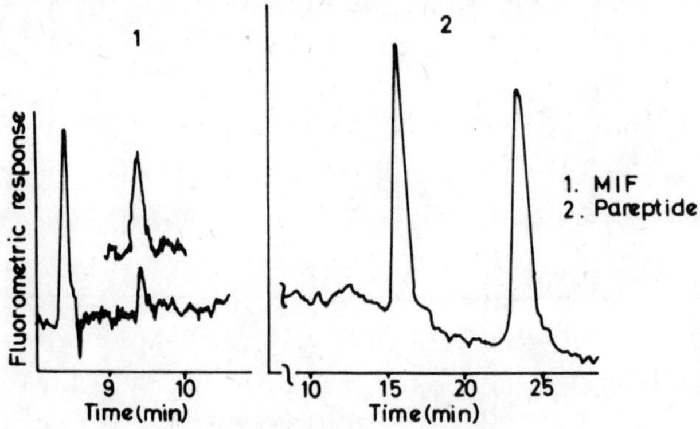

Figure 12 HPLC separation of MIF (4) and Pareptide (B) (26)

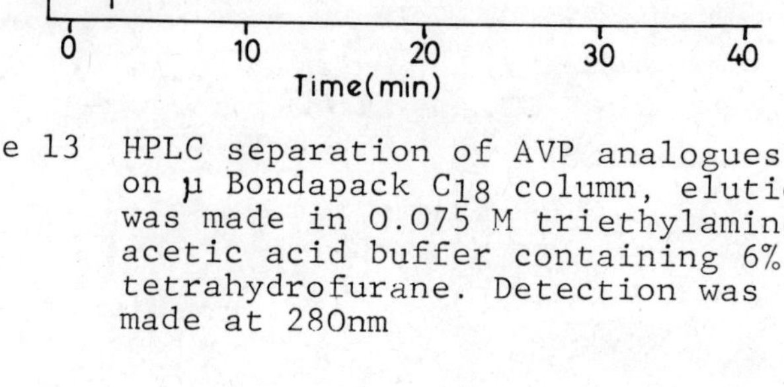

Figure 13 HPLC separation of AVP analogues on μ Bondapack C_{18} column, elution was made in 0.075 M triethylamine - acetic acid buffer containing 6% tetrahydrofurane. Detection was made at 280nm

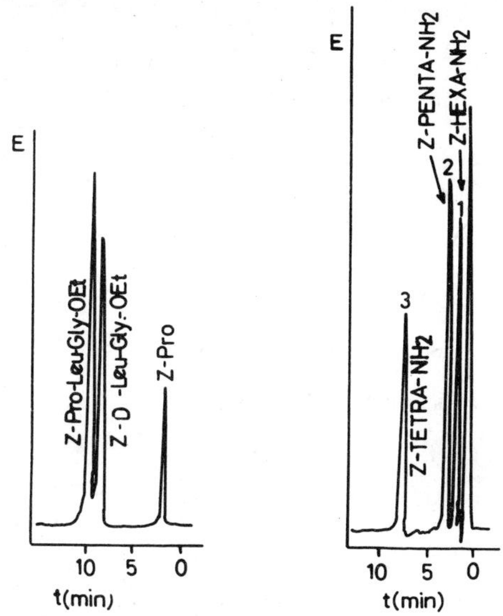

Figure 14 HPLC analysis of several MIF analogues (left side) and fragments (right side) on LiChrosorb RP-8 column (3.2 x 150 nm) using phosphate buffer (pH 7, 0.015 M) and acetonitryl 6:4 with 1 ml/min flow rate (30). Detection was made at 280 nm

phase HPLC (22). Novel data (23-27) indicate HPLC for investigation of OT, VP their analogues and fragments. GLASEL (23) separated the precursors of OT and AVP that is the Neurophysin I and II on reversed-phase high performance liquid chromatography (RP-HPLC) which can be seen in Fig. 9. HRUB and his co-workers also contributed to this topic (24, 25), they successfully separated AVP and its diastereoisomers by RP-HPLC (Figs. 10 and 11). KROL and his co-workers (26) demonstrated the oxytocine C--terminal tripeptide amide fragments (MIF) and one of its structural analogues (pareptide) in blood plasma (Fig. 12).

ADH-specific AVP analogues were synthetized earlier (3-6). The chromatograms of RP-HPLC separation of MIF analogues and fragments are given here (Figs 13 and 14) as two examples of our recent work (28-30).

SUMMARY

An overview is presented here on the chemical aspects, classical liquid chromatography, thin-layer chromatography and high-performance liquid chromatography of oxytocine, vasopressins and several of their analogues and fragments.

ACKNOWLEDGEMENT

The first author of this review spent 15 months in 1972-73 and 6 months in 1981 in Professor M. MANNING's laboratory in the Department of Biochemistry, Medical College of Ohio as an invited research worker. Many thanks again to the leader of the laboratory for the excellent possibilities and successful cooperation.

REFERENCES

(1) WÜNSCH, E. (1974) Synthese von Peptiden. Vol. 15. Houben-Weyl. Methoden der organischen Chemie. Ed. E. Muller, Georg Thieme Verlag, Stuttgart

(2) STEWART, J.M. and YOUNG, J.D. (1969) Solid Phase Peptide Synthesis. W.H. Freeman and Co., San Francisco

(3) MANNING, M., BALÁSPIRI, L., ACOSTA, M. and SAWYER W.H. (1973) J. Med. Chem. 16, 975

(4) SAWYER, W.H., ACOSTA, M., BALÁSPIRI, L., JUDD, J. and MANNING, M. (1974) Endocrinology 94, 1106

(5) MANNING, M., BALÁSPIRI, L., JUDD, J., ACOSTA, M. and SAWYER, W.H. (1974) FEBS Letters 44, 229

(6) MANNING, M., BALÁSPIRI, L., MOEHRING, J., HALDAR, J. and SAWYER, W.H. (1976) J. Med. Chem. 19, 842

(7) BALASPIRI, L., TÓTH, V. M., SIROKMÁN, F., JANÁKY, T. and KOVÁCS, K. (1980) Preparation of ^{14}C- and ^{3}H-labelled vasopressin analogues. Lecture on Annual Meeting of Isotope Work-commettee of HAS, Szeged

(8) TÓTH, V. M., BALASPIRI, L., TAKÁCS, T., KOVÁCS, K. and LÁSZLÓ F.A. (1982) Preparation vasopressin analogues containing unnatural amino acids. Lecture on Annual Meeting of Peptide Work-commettee of HAS, Balatonfüred

(9) BALASPIRI, L., V. TÓTH, M. TAKÁCS, T., TÓTH, G., JANAKY, T., SIROKMÁN, F., KOVÁCS, K. and LÁSZLÓ F.A. Iodination, tritiation and purification of OT, AVP and some of analogues of the latter./In preparation for publication/

(10) LASZLO, F.A., JANÁKY, T., BALÁSPIRI, L. and MORGAT J.L. (1981) J. Endocr. 88, 181

(11) JANÁKY, T., LÁSZLÓ, F.A., BALÁSPIRI, L. and MORGAT, J.L. (1982) Hormon Metab. Res. 14, 369

(12) MANNING, M. (1968) J. Amer. Chem. Soc. 90, 1348

(13) MEIENHOFER, J., TRZECIAK, A., HAVRAN, R.T. and WALTER, R. (1970) J. Amer. Chem. Soc. 92, 7199

(14) TÓTH, V. M. (1982) Synthesis and biological studies of some new vasopressin analogues containing unnatural amino acids. Dissertation, SZOTE, Szeged

(15) TEPLAN, I. (1980) Dissertation of Doctor of Sciences, Budapest, Hungary

(16) LÁSZLÓ, F.A., NAGY, E., GÁSPÁR, L., KÉRI, Gy. and TEPLAN I. (1980) Horm. and Metab. Res. 12, 471

(17) BALÁSPIRI, L. and LÁSZLÓ F.A. Recovery of partly distroyed ^{3}H-LVP by two steps micro-column chromatography. /In preparation for publication./

(18) FRAKER, P.J. and SPECK, J.C. (1978) Biochem. Biophys. Res. Commun, 80, 849

(19) FLOURET, G., TERADA, S., YANG, F., NAKAGAWA, S.H., NAKAHARA, T. and HECHTER, O. (1977) Biochemistry 16, 2119

(20) BALÁSPIRI, L., TÓTH, G. and KOVÁCS, K. (1982) Synthetic aspects and highlights of structure-activity relationships in vasopressins. In Proc. of 9th Hungarian Endocrin. Congress, Ed. F.A. LÁSZLÓ pp. 95-102

(21) MANNING, M., GZRONKA, Z. and SAWYER W.H. (1981) The Pituitary. Eds. C. Beardwell and G. Robinson, Butterworths, Kent England pp. 265-296
(22) BLEVINS, D.D.,BURKE, M.F. and V.J. HRUBY (1980) Anal. Chem. 52, 420
(23) GLASEL, J.A. (1978) J, Chromatography 145, 469
(24) LARSEN, B., FOX, B.L., BURKE, M.T., and HRUBY, V.J.(1979) Int. J. Peptide Protein Res. 13, 12
(25) BLEVINS, D.D., BURKE, M.F., HRUBY, V.J. and LARSEN, B.R. (1980) J. Liquid Chromatography 3, 1299
(26) KROL, G.J., BANOVSKY, J.M., MANNAN, C.A., PICKERING, R.E. and KHO, B.T. (1979) J. Chromatography 163, 383
(27) NACHTMANN, F. (1979) J. Chromatography 176, 391
(28) BALÁSPIRI, L. and OHMACHT, R. Purification and purity control of some AVP analogues by reversed-phase High Performance Liquid Chromatography. /In preparation for publication./
(29) BALÁSPIRI, L. and SZÓKÁNY Gy. Purification and purity control of synthesis MIF analogues by reversed-phase High Performance Liquid Chromatography. /In preparation for publication./
(30) BALÁSPIRI, L., V. TÓTH, M., TAKÁCS, T., KOVÁCS, G.L., TELEGDY, Gy. and KOVÁCS K. Synthesis and Biological data of Pro-Leu-Gly-NH_2 /MIF/ analogues. /In preparation for publication./

Proc. of the Symp. on Advances in Liquid Chromatogr., Szeged, Hungary, 1982

MICRO COLUMN CHROMATOGRAPHY OF LARGE PEPTIDES

E.S. GANKINA, I.O. KOSTIUK, B.G. BELENKII

Institute of Macromolecular Compounds,
Academy of Sciences of the USSR,
Leningrad, Soviet Union

INTRODUCTION

The application of micro columns for liquid chromatography was first put into practice by Sandakhchiev, Grachev and Kuzmin (2). Horváth and Lipsky (1) in 1967 used micro column with 0.5 mm in diameter, packed with pellicular ion exchanger. Subsequently, separation of substances on small bore columns has been developed by several groups of scientists (3-6).

The advantages of micro column chromatography (MCC) are its capability to provide high efficiency, high speed and high mass sensitivity which renders the adequate analysis of substances on picogram range. An additional factor to be mentioned is that the operating cost is reduced by a factor of nearly 100 to 1000 including the economy of both the sorbent and the eluent.

Micro columns for gel chromatography were reported mainly for the group separation of iodinated compounds (7) and the micro bore column chromatography using long capillary tubes rendered high efficiency besides being economical with the sorbent, eluent and substances to be separated. As early as in 1966, Nystrom and Sjövall (8) wrote the separation of vitamin K analogues (about 20 μg each of vitamins $K_{2(40)}$-$K_{2(10)}$) on a capillary column of methylated Sephadex G-25 superfine where the internal diameter of tubing was 1.5 mm.

The gel chromatographic separation of peptides of similar molecular sizes is based on the use of high selectivity processes. Either the HPLC sorbents for gel chromatography (9-11) can be used or the increase of efficiency by the application of cascade or multidimensional chromatographic processes (12-13) can be performed or the combination of these two, above mentioned methods, that is the cascade or multidimensional system of HPLC columns can be used.

Conventional chromatography uses a system of valves for controlling the migration of substances along a system of separating columns. However, this approach is quite unsuitable for MCC because of its considerable extra column band spreading.

The present work describes the application of cascade of columns to MCC. Columns consist of polyethylene capillaries which can easily be connected to each other or to the outlet tube of the detector (Fig.1). This procedure of carrying out cascade MCC can virtually completely prevent zone spreading in column connections.

MATERIAL AND METHODS

Materials:

Cytochrome C (Schuchardt, FRG), insulin (Riga Meat Plant, SU), freshly distilled cyanogen bromide (Serva, FRG) formic acid C.P. grade (Soyuzreaktiv,SU), guanidine hydrochloride, pure grade (Soyuzreaktiv,SU), Sephadex G-10, Sephadex G-50,superfine (Pharmacia Fine Chemicals Uppsala, Sweden), Spheron P-1000 (Lachema, Czechoslovakia).

Apparatus:

A Kh Zh-1305 micro column liquid chromatograph (Special Design Bureau of Analytical Instruments, Academy of Sciences of the USSR) and a spectrophotometric detector with a 1 ml flow cell were used.

Methods:

Polyethylene columns with i.d. of 0.085 to 1.00 mm and 22 and 11 cm in length were used. The columns were packed with a 80 % slurry of Sephadex G-50 (20-30 μm fraction in the wet state) swollen in the elution buffer. The sorbent slurry was pumped into a capillary container 20-25 cm in length, the column was connected with it and the sorbent slurry was introduced into it with an eluent stream at a flow rate of 250 μl/h. When the sorbent level reached the column edge, the eluent was pumped

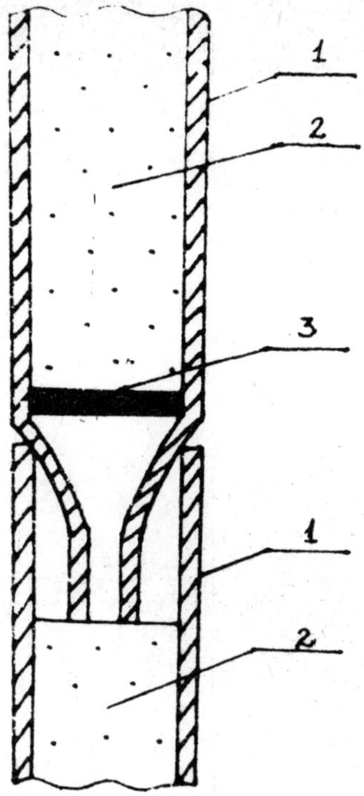

Fig.1. Scheme of connections of micro columns in cascade MCC. 1: Polyethylene micro column. 2: Packings, 3: Porous titanium filter.

through the whole system at a rate of 500 µl/h until the sorbent level ceased to decrease. Subsequently the capillary container was disconnected from the system and the column was washed with the flow rate of eluent of 250 µl/h until a stable baseline was established.

The columns were packed with Spheron P-1000 (20-30 um fraction in wet state) in a similar manner by using a 50 % slurry in an ethanol-water mixture (9:1) by volume, the flow rate of eluent was 5.4 ml/h. The column was washed with distilled water (2-3 ml/h) and equilibrated with the eluting buffer at a flow rate of 2 ml/h.

The operation proceeded as follows: the components that had to be additionally separated after undergoing chromatography on one column or on several columns connected in series were introduced into a column directly connected to the outlet of the detector of the micro column chromatograph. Subsequently this

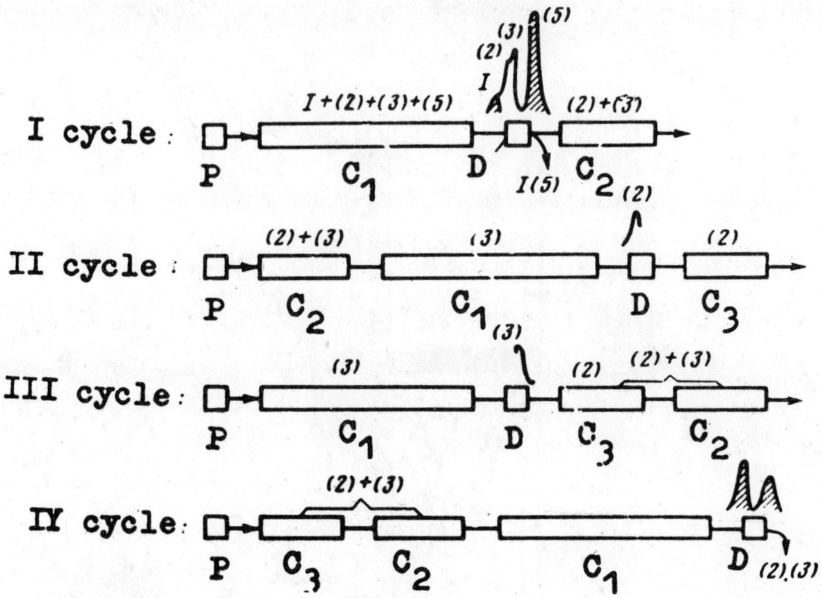

Fig.2. Scheme for recycling exclusion MCC of a mixture of 1) initial insulin, 2) insulin chain B, 3) insulin chain A and 5) low molecular weight products, P: pump, D: detector, C_1, C_2 and C_3: columns.

Curve above detector: recording of chromatograms in each cycle (shaded peaks represent collected fractions).

column was disconnected from the detector and connected to a new system of columns or was directly eluted with a new solvent by placing it between the pump and the detector. These operations can be carried out either on one micro column chromatograph or several chromatographs can be used simultaneously, i.e. the column-detector system can be regarded as a separating unit as it is usually done in cascade chromatography (14).

RESULTS AND DISCUSSION

The reaction mixture of denaturated, reduced and carboxymethylated insulin (mixture of A and B insulin chain (16) was separated on three polyethylene columns, 0.085 cm in diameter and 28, 12 and 11 cm in length packed with Sephadex G-50 in a 5 M guanidine hydrochloride aqueous solution. The separation scheme is shown in Fig.2. One column (C_1) was stationary and

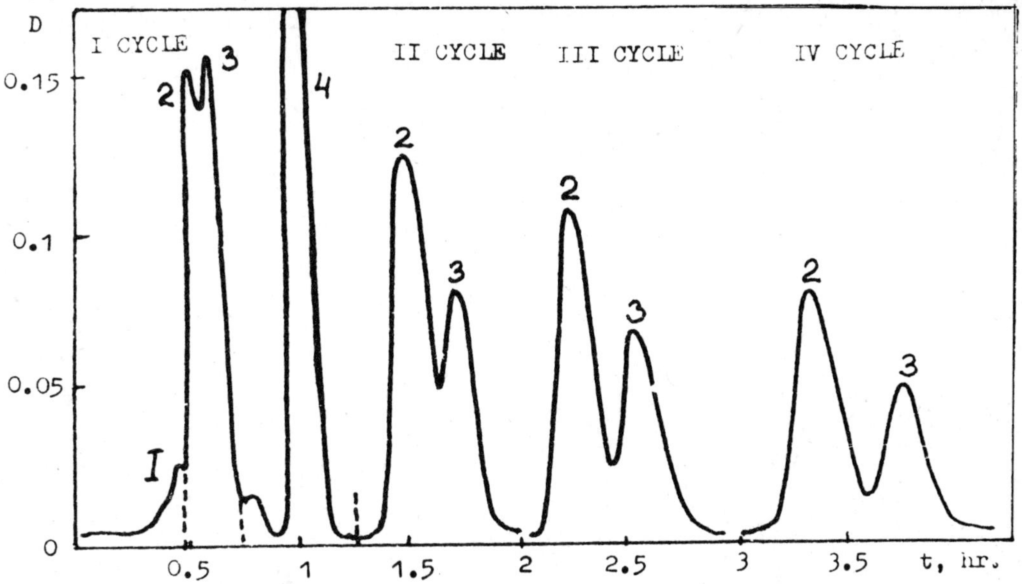

Fig.3. Recycling exclusion MCC of 1.5 nmol of the reaction mixture of insulin on three columns (0.085 x 28, 12 and 11 cm) packed with Sephadex G-50, eluent: 5 M guanidine hydrochloride solution, elution rate: 0.16 ml/h.

was connected to the detector cell, the other two columns were replaceable (C_2 and C_3). One of the replaceable columns (C_2) was connected to the detector outlet and served as a receiver for the separated components. After the required fraction had been transferred into C_2 under detector control, it was disconnected and connected in front of C_1, directly to the injector. Thus, the succeeding separation cycle was carried out on two columns (C_1 an C_2). If the fraction directed to recycle was of large volume, two columns, C_2 and C_3 were placed in series after the detector and then connected in front of the stationary column for the next separation cycle. Fig.3 shows the elution profile of the separation.

It is clear that the complete separation of A and B insulin chains proceeds in four cycles (about 4 hours). The separation time of this mixture by conventional recycling chromatography was about 40 hours (16). However, if 5 M guanidine hydrochloride is used as eluent, the fractions should be desalted for the

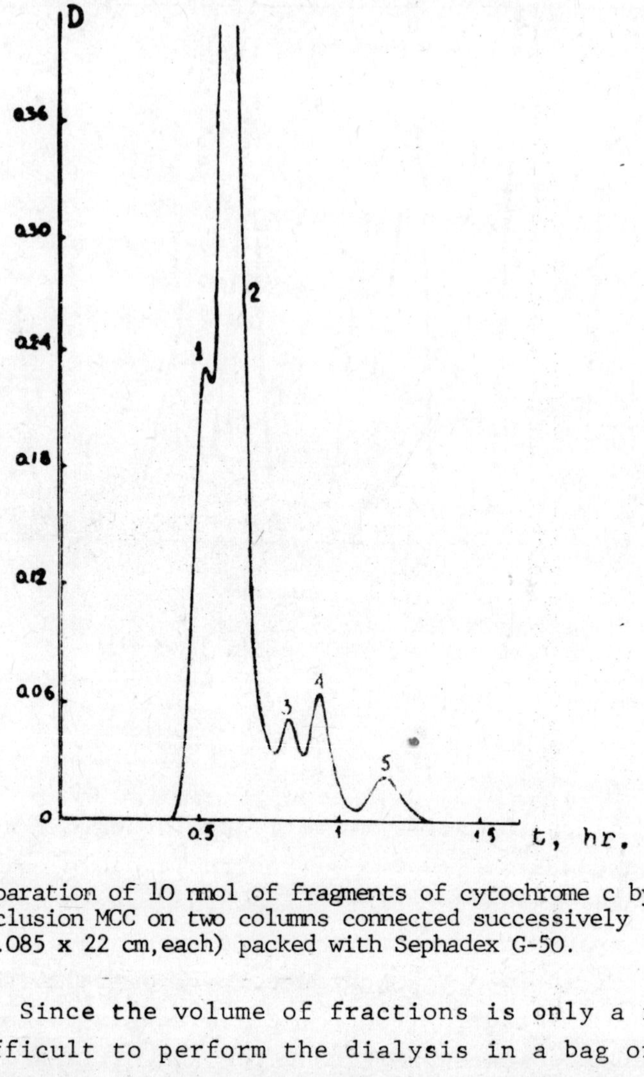

Fig.4. Separation of 10 nmol of fragments of cytochrome c by exclusion MCC on two columns connected successively (0.085 x 22 cm, each) packed with Sephadex G-50.

further work. Since the volume of fractions is only a few ml, it is rather difficult to perform the dialysis in a bag or other vessels and the possible loss can also be considerable. Cascade MCC makes it possible to carry out desalting of fractions directly in the course of MCC with a minimum loss of sample. The required fractions were transferred into a column with i.d. of 1 mm and 21.5 cm in length packed with Sephadex G-10, equilibrated with an ammonia solution (pH 9) and attached to the cell outlet. Subsequently, elution was carried out with the same buffer, flow rate was 1 ml/h and the first peptide peak was collected in an ampule. The eluate was freeze-dried and used for further investigation.

The product of partial cyanogen bromide digestion of cytochrome C was obtained according to ref.(17). It contains a mixture of five peptides, characterized by their order, relative molecular weight and number of amino acid residue comprising the peptides 1) 8000 and 1-80; 2) 6500 and 1-65; 3) 3800 and 66-104; 4) 2300 and 81-104; 5) 1400 and 66-80, respectively.

These peptides have been completely separated (17) by gel chromatography on Sephadex G-50 gel column (5 x 250 cm) using 10 % aqueous formic acid solution as eluent, in 72 hours. By MCC, the partial cyanogen bromide digest of cytochrome c was separated on four columns. C_1 and C_2 were packed with Sephadex G-50, each 22 cm in length, C_3 and C_4, each 11 cm in length, were packed with Spheron P-1000. The peptide mixture was separated into four fractions on two columns, C_1 and C_2, connected in series. A 10 % aqueous formic acid solution was used as eluent (Fig.4), the first fraction was a mixture of poorly separated peptides 1) and 2). This fraction was injected onto the column C_3. The peptides 3) and 4) were separated by the method of recycling in micro column and the fractions were collected after the third cycle. Peptide 5) was well separated and collected at once. For complete separation of peptides 1) and 2), columns C_1 and C_2 were replaced by columns C_3 and C_4. 10 % aqueous formic acid solution containing 3 % of tert. butanol was used as eluent. First of all, column C_4 was connected to the detector and washed with the eluent while the base line became stable. Subsequently, column C_3 was connected to it from the above and thus peptides 1) and 2) could be introduced into the system. These peptides were completely separated during one cycle at an elution rate of 500 µl/h. The elution profile for the complete separation of cytochrome c fragments is shown in Fig.5. The separation time was only about 3 hours, i.e. the procedure of separation was 18 times faster than for the usual process (17).

In the above case of separations different packings in columns (Sephadex and Spheron) and recycling had to be used to reach the adequate resolution. However, to improve the efficiency of the separation system, multidimensional processes involving not only gel chromatography but adsorption or reverse phase

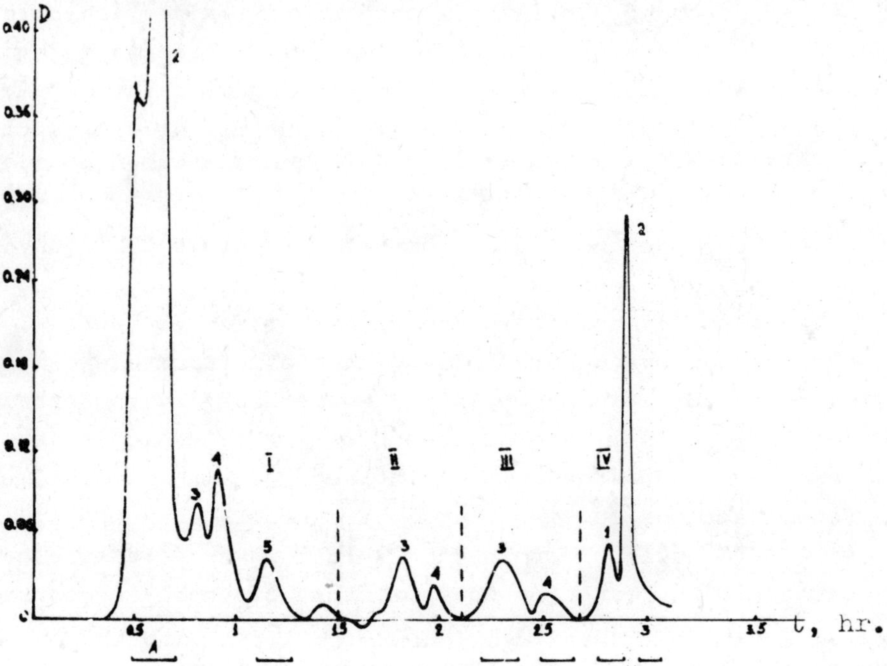

Fig.5. Cascade MCC of 20 nmol fragments of cytochrome c on two columns (0.085 x 22 cm) packed with Sephadex G-50 (eluent 10 % aqueous formic acid solution with 250 ml/h flow rate) and two columns (0.085 x 11 cm) packed with Spheron P-1000 (eluent 10 % aqueous formic acid containing 3 % of tert.butanol with 2 ml/h flow rate) Dotted line shows limits between separation cycles on Sephadex. Cycle lV represents the separation of fraction A on Spheron P-1000. Bold line shows fractions separated in each cycle.

chromatography, too, can also be used in the cases of peptides with similar molecular weights.

If the peptides are detected at 280 nm, the sensitivity of monitoring is determined by the aromatic amino acid residues of the solutes. The analytical recycling MCC if insulin fragments was carried out with 1.0 to 1.5 nanomol of the sample and the peak height was 80 % and 20 to 30 % of the full scale in the first cycle and in the fourth cycle, respectively. For the separation of cyanogen bromide digest of cytochrom c, the analytical possibilities were about 10 nanomol because of different amounts of aromatic amino acid residues in the peptides, the peak heights were ranged from 80 % to 2.5 % of the full scale.

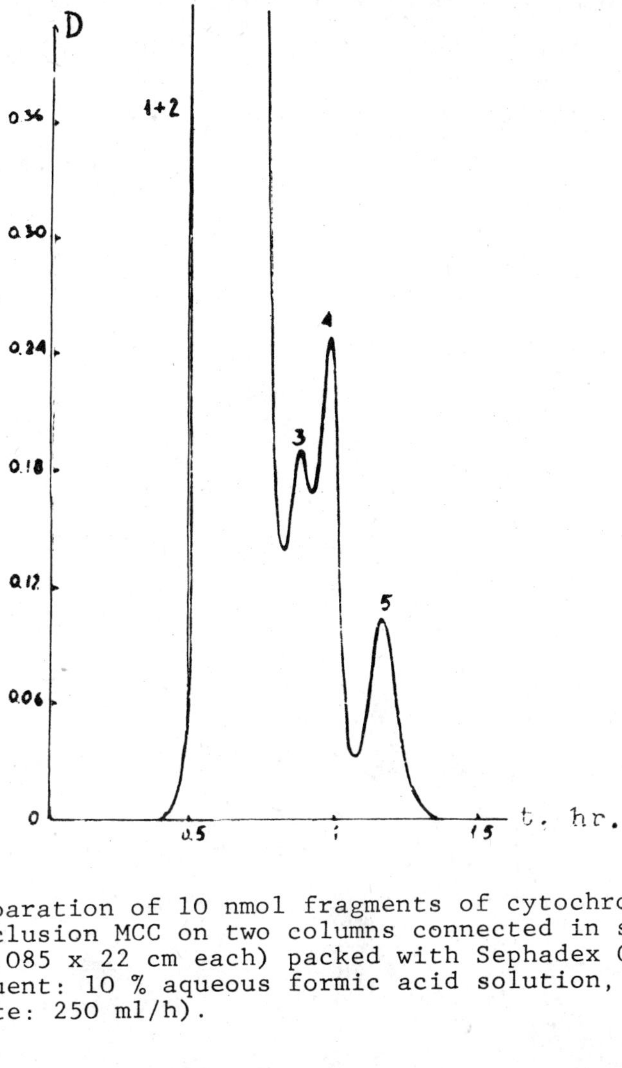

Fig.6. Separation of 10 nmol fragments of cytochrome c by exclusion MCC on two columns connected in series (0.085 x 22 cm each) packed with Sephadex G-50. Eluent: 10 % aqueous formic acid solution, elution rate: 250 ml/h).

Cascade MCC enables detecting from 1 to 10 nanomol of peptides after three or four separation cycles.

Fig. 6 shows the elution profile in the case of high overload of MCC relative to the above described experiments.The separation of 60 nanomol of cytochrom c fragments on a Sephadex G-50 gel column is demonstrated using 10 % aqueous formic acid solution as eluent. The comparison of Figs 4 and 6 shows that the six fold increase of the amount of peptide does not lead to any appreciable decrease of the separation.

SUMMARY

The method of exclusion micro column chromatography of large peptides has been described. Polyethylene columns with inside diameter of about 1 mm, directly connected to each other were used, the examples of separations are the reaction mixture of denatured reduced and carboxymethylated insulin as well as the products of partial digestion of cytochrome c.

This procedure prevents zone spreading in consequence of the sample valwes when columns with the same or different packings are connected directly in series for the transfer of the total sample or its fractions.

The method can be used for both separation with recycling and the desalting of the eluates.

Sephadex G-50 and Spheron columns were used. The suggested type of exclusion micro column chromatography can be applied for the separation of up to 60 nmol of large peptides.

REFERENCES

(1) HORVÁŢH,Cs. and LIPSKY,S.R. /1967/ Anal.Chem. 39 1422
(2) KNORRE,D.G. and VENKSTERN,T.V.(Ed.) Ultramicroanalysis nukleinovykh kislot. Moscow, Edition Nauka, 1973.
(3) ISHII,D., ASAI,K., HIBI,K., JONOKUCHI,T. and NAGAYA,M. /1977/ J.Chromatogr. 144 157
(4) SCOTT,R.P.W. and KUCERA,P. /1979/ J.Chromatogr. 169 51
(5) TSUDA,T. and NOVOTNY,M. /1978/ Anal.Chem. 50 271
(6) KEVER,E.E., GANKINA,E.S. and BELENKII,B.G. /1981/ Vysokomol. Soedin.Ser. A-23 234
(7) FISCHER,L.: An Introduction to Gel Chromatography, P.281. North Holland, Amsterdam, 1971
(8) NYSTRÖM,E. and SJÖVALL,J. /1966/ J. Chromatogr. 24 212
(9) UI,N.J. /1981/ J.Chromatogr. 215 289
(10) REGNIER,F.E. and GOODING,K.M. /1980/ Anal.Biochem. 103 1
(11) ROUMELIOTIS,P. and UNGER,K.K. /1979/ J.Chromatogr. 185 445
(12) MAJORS,R.E. /1980/ J.Chromatogr. Sci. 18 571
(13) FREEMAN,D.H. /1981/ Anal.Chem. 53 2
(14) GANKINA,E.S., KOSTUIK,I.O. and BELENKII,B.G. /1979/ Bioorganich.Khim. 5 325
(15) WILGUS,H. and STELLWAGEN,E. /1979/ Anal.Biochem. 94 228

Proc. of the Symp. on Advances in Liquid Chromatogr., Szeged, Hungary, 1982

SEPARATION OF PROTEIN-DEPRENYL ADDUCTS BY GEL CHROMATOGRAPHY

É. SZÖKŐ,* H. KALÁSZ,** K. MAGYAR*

*Department of Pharmacodynamics,
**Department of Pharmacology, Semmelweis University of Medicine, Budapest, Hungary

SUMMARY

The very strong or irreversible deprenyl binding of serum proteins was proved by gel chromatography. Separation of deprenyl containing protein fractions indicated that the largest amount of ^{14}C-deprenyl occurs in the macro-globulin fraction while the albumin kept the smallest portion of the compound. The binding remained unaltered during the several hours of gel chromatography and freeze drying. The serum albumin-deprenyl adduct was identified by immuno-electrophoresis while the characterization of the other binding proteins is in progress.

INTRODUCTION

(-) Deprenyl, N-methyl-N-propargyl-(2-phenyl-1-methyl)-ethyl-ammonium HCl, JUMEX^R (Chinoin, Budapest, Hungary) is a highly potent, selective inhibitor of the B-type monoamine oxidase (1,2). Deprenyl has been successively used in the therapy of Parkinson's disease as it strongly reduces the dose of L-DOPA without the so called "cheese effect" which often occurs on administration of several monoamino oxidase inhibitors.

Deprenyl was ^{14}C -labelled at position 2 of the ethyl-2-phenyl radical as shown in Fig. 1.

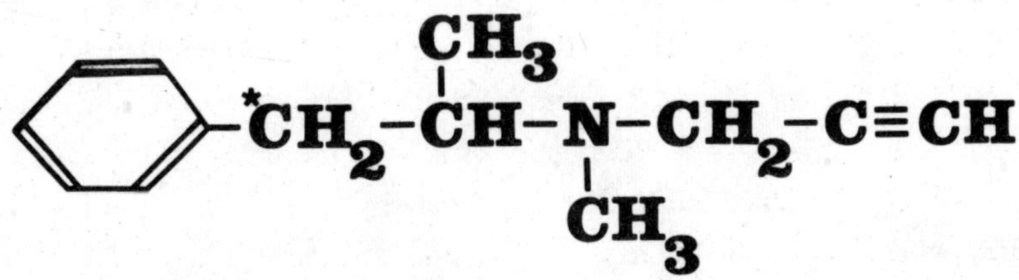

Figure 1 The structure of deprenyl.

The fate of deprenyl in the body, including the metabolism (3-5) and localisation (2) of the compounds as well as its binding (6,7) were thoroughly studied, but little was known about the nature of the adduct.

Plessas (6) and Benakis (7) described the in vitro and in vivo binding of deprenyl. According to their reports, the binding between rat serum and deprenyl was determined after incubation, centrifugation and ultrafiltration. They found that 75% and 85% of deprenyl was bound to the plasma when the incubated amount was 0.041 and 0.15 μg/ml, respectively.

RESULTS AND DISCUSSION

The aim of our investigation was to study the deprenyl binding capacity of the human serum and the nature of the adducts. At the beginning of our research, the existence of the binding was confirmed by the method of equilibrium dialysis, the majority of the radioactivity moved through the semipermeable membrane to the cell containing the serum (8). These results were found both with human and rat serum while the previous freeze drying destroyed its deprenyl binding capability.

To perform deprenyl-protein binding, serum was incubated with radiolabelled deprenyl for 2 hours. The sample was then immediately separated by gel chromatography using Sephadex G-15 column having 2.5 cm in diameter and 90 cm in length and eluting with 0.05 M ammonium acetate aqueous solution as mobile phase.

The deprenyl-serum samples were sometimes kept at -25°C for a few days but this delay has never changed the ratio of the bound and free deprenyl.

The deprenyl-protein adduct(s) eluted at the void volume of the column as the molecular weight of any protein adduct highly exceeded the exclusion limit of the Sephadex G-15 which is specified as 1500 (9).

The gel chromatographic pattern of such an experiment is given in Fig. 2.

The fractions eluted at the void volume were collected together. The further separation was performed on Sephadex G-200 gel column (5 cm x 60 cm), the gel chromatogram (Fig. 3) indicated three peaks representing mainly macro-globulins, gamma globulins and albumin, but numerous other proteins can be also found there. The most striking characteristic of Fig. 3 is that the large portion of radioactivity is shown by the first eluting peak, together with the slightest UV extinction. At the same time, the albumin peak (the third one) possesses the smallest amount of radioactivity but the highest UV extinction. The chromatogram did not indicate any amount of free deprenyl, that is the dislocation or dissociation of bound deprenyl (which

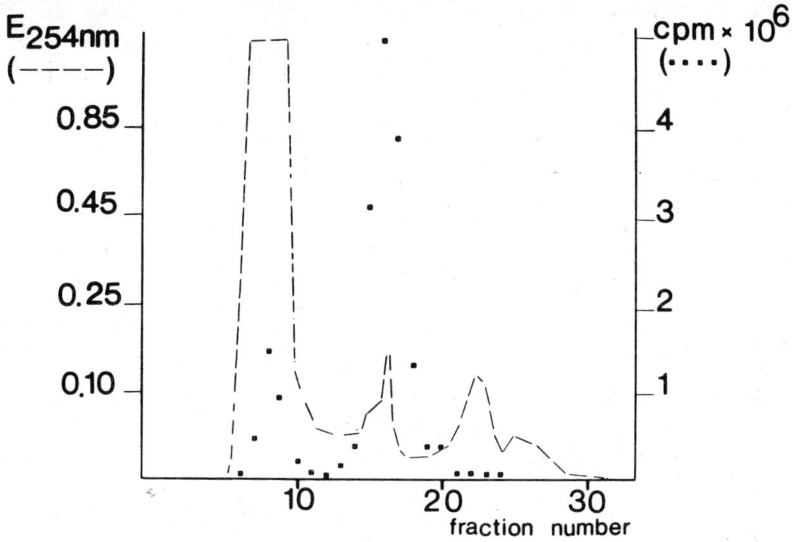

Figure 2. Gel chromatography of free and bound deprenyl.

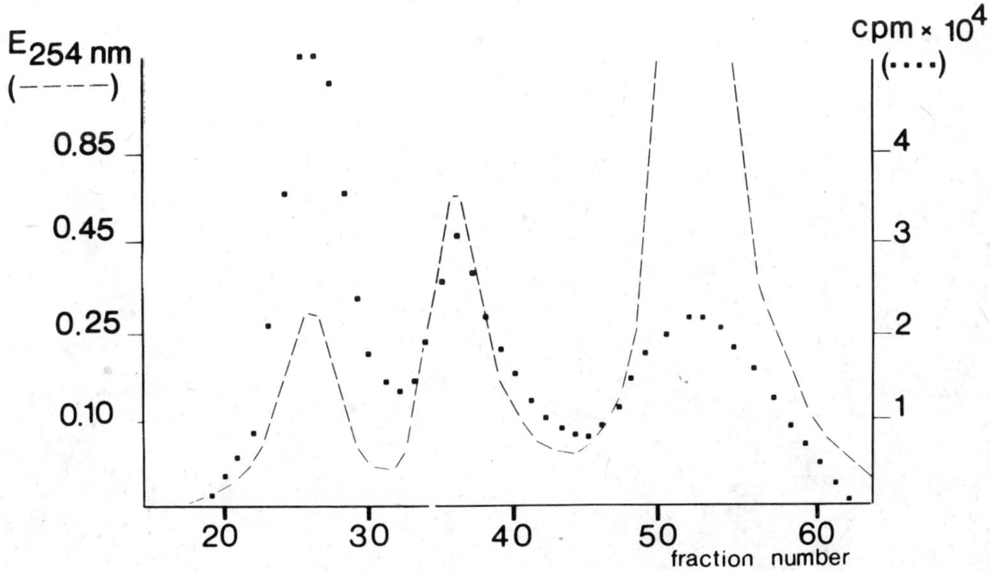

Figure 3. Separation proteins containing deprenyl.

may be eluted at the salt fraction or even later) and this fact definitely supports the strong or irreversible binding of deprenyl to the serum proteins.

Although the separation detailed above means only a partial purification of the deprenyl-protein adducts, that is several proteins were collected in the same fractions, we wished to characterize the deprenyl binding molecules. As the amount of the adducts was small, immunoelectrophoresis became the method of choice. For the time being, we completed the identification of the third eluting peak with the conclusion that albumin itself is the protein which is solely responsible for deprenyl binding in this group.

The identification of the protein part of the other adducts is in progress.

CONCLUSIONS

Gel chromatography of the in vitro incubated deprenyl protein mixture showed well defined peaks which serve not only the determination of bound deprenyl but also supports its strong binding character.

Further separation on Sephadex G-200 differentiated three well defined peaks with various UV extinction - radioactivity ratios and rendered additional prove of the stability of the adducts.

REFERENCES

(1) KNOLL,J. and MAGYAR,K. Adv. Biochem. Pharmacol. 5: 393 (1972)

(2) MAGYAR,K., SKOLNIK,J. and KNOLL,J. In: V. Conf. Hung. pro Therapia et Investigatione in Pharmacologia. (Ed:Leszkovszky,G.) p. 103, Akad. Kiadó, Budapest, (1968)

(3) REYNOLDS,G.P., ELSWORTH,J.D., BLAU,K., SANDLER,M., LEES,J.A. and STERN, G.M. Br.J.Cl.Pharmacol. 6, 542, (1978)

(4) KALÁSZ,H. Journal of High Resolution Chromatography & Chromatography Communications, 6, 49 (1983)

(5) KALÁSZ,H., NAGY,J. and KNOLL,J. In: Chromatography and Mass Spectrometry in Biomedical Sciences 2(Ed:Frigerio,A.) p.203, Elsevier, Amsterdam, (1983)

(6) PLESSAS,Ch.: non published data

(7) BENAKIS,A. Abstr. of 12th CINP Congress. Stockholm, Sweden, (1980)

(8) SZÖKÖ,E., KALÁSZ,H. and MAGYAR,K. Abstr. of Symp. Hung. Polish Pharm. Soc. Visegrad, Hungary, (1983)

(9) KALÁSZ,H.: Gel Chromatography In: Protein Investigation Methods. (Ed:Kerese,I.) Ellis Harwood - Akademiai Kiadó, Chichester - Budapest, (1983)

Proc. of the Symp. on Advances in Liquid Chromatogr., Szeged, Hungary, 1982

PURIFICATION OF SUPEROXIDE DISMUTASES FROM DIFFERENT SOURCES

B. MATKOVICS, L. SZABÓ

Biological Isotope Laboratory, József Attila University, Szeged, Hungary

SUMMARY

Results indicate that polyacrylamide gel electrophoresis (PAGE) can be used for checking the purity of enzyme preparations as well as for the separation of superoxide dismutases from homogenates. PAGE is also the method of choice for detection of isoenzymes of superoxide dismutases. At the same time, the comparative quantitative amino acid analysis can yield very important data on the integrity of enzyme protein if the hydrolysis and the whole procedure have taken place under identical conditions.

INTRODUCTION

Superoxide dismutases (SODs) are extremely unusual enzymes because their substances are free radicals (4, 5, 15), namely $/O_2^-/$ radicals (11, 15). These enzymes occur in all aerobic cells and they are generally thought to be essential for aerobic life. Superoxide dismutase has found increasing usage in biochemistry, radiobiology and medicine. In most cases Cu,-Zn-SOD from bovine erythrocytes was used. The standard extraction procedure utilizes a precipitation step by the mixture of chloroform and ethanol but it has several disadvantages. To eliminate the problems of the chloroform-ethanol precipitation we elaborated a combination of several purification steps (18).

The first protein showing SOD activity was erythrocuprein (10, 14), a copper-zinc-containing enzyme which has recently been isolated from cytosols of a wide range of eukaryotic organisms. Prokaryotes have manganese- or iron-containing SOD, these enzymes can not be inhibited by cyanide which means a sharp contrast to the Cu,-Zn-SOD.

Numerous papers have been published concerning the purification and amino acid analysis of SODs (2, 4, 19) but they mainly report the particular steps of the separations. This paper intends to give a comprehensive evaluation of polyacrylamide gel electrophoresis and amino acid analysis of SOD enzymes.

METHODS

The hydrolysis of enzyme proteins was carried out by 6 N hydrochloric acid at 105°C for 48 hours.

The amino acid analysis was performed by BioCal BC 200 automatic amino acid analyser according to the method published by MOORE et al (17), HANNIG (7), HAMILTON (6) and modified by KERESE et al. (13).

The total SOD activity was determined by the method of MISRA et al. (16) which is based on the inhibition of the epinephrine-adrenochrome superoxide /O_2^-/ dependent transformation. Mn-SOD was determined by adding 2 mM KCN to the measuring solution. Then, the measurable SOD activity was considered, in good agreement with the inventors of the method, as the quantitative value of the mitochondrial Mn-SOD. The data of these activities are shown in Fig. 11.

The PAGE separation and demonstration of the SODs were described by BEAUCHAMP et al (2). They used p-nitroblue tetrazolium chloride (NBT) for the demonstration of the enzyme, as the site of the SOD appears as a colourless streak on the blue-stained gel (HARRIS, 9). On the other hand, xanthine oxidase shows a positive reaction and appears as intensive blue bands (Fig. 10).

The protein staining of the gel was performed by the use of Comassie Brilliant Blue R-250 or Amido Black applying their solution in 2 %, the time period of staining was 20 minutes.

The apparatus used for PAGE separation was purchased from LKB Produkter (Bromma, Sweden), buffers and conditions were identical with those described by DAVIS et al. (3).

Producer of the Cu,-Zn-SOD was ORGOTEIN (Diagnostic Data Labor, Inc., Palo Alto, USA) (12), bovine and canine red blood cell SODs were obtained from Sigma (Saint Louis, USA). Any other SODs were purified in our Laboratory with the use of Sephadex G-75 or Sephadex G-100 gel columns, the typical example of separations is demonstrated in Fig. 6.

RESULTS

Fig. 1 shows the amino acid analysis of Cu,-Zn-SOD purified from human red cells (RBC) when isolation was derived from normal RBC (obtained from a blood transfusion unit) or from RBC of diabetic patients (8). We suppose that the lower values of the His and Cys contents are mainly indicative regarding the oxidative damage of the SOD molecules in diabetics.

Fig. 2 demonstrates the data of our own amino acid analysis of SOD isolated from normal RBC, and compared with the data of HARTZ et al. (10) who purified SOD from human sources using different methods. There is a significant difference between the two sets of data.

Column 3 of Fig. 3 shows the amino acid analysis of SOD originated from a similarly normal human RBC (14) and the calculated values, which make the checking and comparison easier.

Fig. 4 shows the results of the amino acid analysis of vegetable, mainly leguminous, Cu,-Zn-SOD (partly our own data, as HANUSZ (8), partly literary data as ASADA et al. (1), SAwada et al. (19).

The first line of Fig. 5 shows a quantitative comparison of human placenta and RBC Cu-,Zn-SOD (erythrocuprein) and cerebrocuprein (cat brain) amino acids according to the results of HANUSZ (8) and PHAM VAN HIEN et al. (18).

We conclude that comparative quantitative amino acid analysis can yield very important data concerning the state of the enzyme protein, and that the data of hydrolysis and amino acid analysis carried out under identical circumstances are easy to compare.

Amino acid	Calculated (assumed) number of amino acids of normal RBC Cu,-Zn-SOD per 33.600g of protein	Calculated number of amino acid residues of diabetics RBC Cu,-Zn-SOD per 33.600g of protein
Lys	18	18
His	14	9
Arg	10	18
Asp	33	34
Thr	22	20
Ser	26	23
Glu	21	19
Pro	23	18
Gly	50	46
Ala	20	19
Val	22	22
Met	8	10
Ileu	16	15
Leu	14	19
Tyr	10	20
Phe	11	18
Trp	5	5
Cys/half	8	3

Figure 1 Comparison of amino acid composition of normal and diabetic human red blood cell Cu,-Zn-SOD (Amino acid content mole/mole enzyme)

Amino acid	Assumed number of human (h.) RBC SOD amino acid residues per 33.600 g of protein (Hanusz, 1974)	Assumed number of h. RBC amino acid residues/ 33.600g of protein (Hartz et al. 1969)
Lys	18	22
His	14	16
Arg	10	7
Asp	33	36
Thr	22	16
Ser	26	20
Glu	21	26
Pro	23	10
Gly	50	50
Ala	20	20
Val	22	28
Met	8	0
Ileu	16	16
Leu	14	17
Tyr	10	0
Phe	11	8
Trp	5	–
Cys/half	8	7

Figure 2 Comparison of assumed amino contents of normal human RBC Cu,-Zn-SOD /literary and our own data/

Amino acid	(see Fig. 2. also about first and second columns) 1.	2.	Assumed number of h. RBC SOD amino acid residues (Markowitz et al. 1959) 3.	Calculated number of h. SOD amino acid residues m.w. of 33.200 4.
Lys	18	22	20	20
His	14	16	14	14
Arg	10	7	9	9
Asp	33	36	42	41
Thr	22	16	18	18
Ser	26	20	20	20
Glu	21	26	28	28
Pro	23	10	11	11
Gly	50	50	46	46
Ala	20	20	21	21
Val	22	28	23	23
Met	8	0	–	–
Ileu	16	16	11	11
Leu	14	17	18	18
Tyr	10	0	2	2
Phe	11	8	10	10
Trp	5	–	–	1
Cys/half	8	7	11	11

Figure 3 Amino acid composition of human RBC Cu,-Zn-SOD /Calculated and measured values/

Amino acid	Cu,-Zn-SOD from green pea, spinach and different beans				
	1.	2. (lit.)	3. (lit.)	4.	5.
Lys	20	10	13	21	27
His	15	18	14	12	10
Arg	21	6	7	16	16
Asp	25	45	35	38	33
Thr	16	30	28	15	15
Ser	16	14	10	27	30
Glu	18	19	20	27	22
Pro	26	14	17	13	12
Gly	28	56	42	45	48
Ala	26	21	23	13	14
Val	22	21	28	16	20
Met	7	0	2	28	16
Ileu	19	20	6	8	9
Leu	20	21	22	19	18
Tyr	14	0	0	13	6
Phe	8	9	6	10	8
Trp	4	0	0	5	5
Cys/half	8	6	4	5	5

Figure 4 The amino acid contents of the purified Cu,-Zn-SOD from spinach and Leguminosae /literary and our own examinations/
/SAWADA et al. 1972; ASADA et al. 1973; HANUSZ 1974/

Amino acid	1.	2.	3.
Lys	21	18	25
His	7	14	16
Arg	8	10	16
Asp	31	33	26
Thr	19	22	11
Ser	21	26	14
Glu	50	21	25
Pro	20	23	22
Gly	23	50	25
Ala	29	20	25
Val	26	22	24
Met	3	8	16
Ileu	8	16	30
Leu	25	14	11
Tyr	5	10	7
Phe	12	11	10
Trp	0	5	5
Cys/half	2	8	8

Figure 5. Comparison of the amino acid content of human placenta SOD (first line), RBC Cu-, Zn-SOD (second line) and cerebrocuprein (third line).

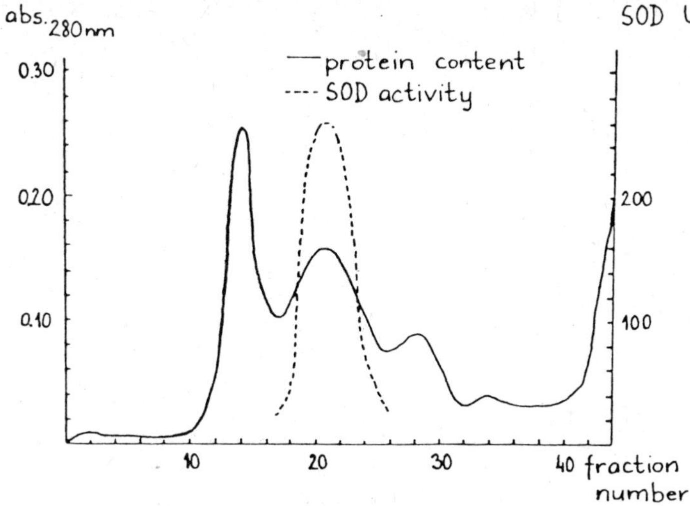

Figure 6 Elution pattern of Cu,-Zn-SOD from Sephadex G 75 column

In the following we give a picture of the purification of SOD and the PAGE comparative analysis of SODs of different origins.

Here, too, we call special attention to the methods we used. Fig. 6 shows the purification of erythrocuprein on a Sephadex G-75 column when elution was made at 0.1 M KCl at a flow rate of about 35 ml/hr.

In Fig. 7 we show the electrophoretogram of beef liver erythrocuprein (Orgotein^{R}) applied in three concentrations. It can be well seen that in higher concentrations the more slowly crawling streaks of Mn-SOD also appear beside the streaks of the middle erythrocuprein.

Figure 8 shows the steps of erythrocuprein and Mn-SOD enrichment using rat liver. The columns represent the specimens applied for steps of purification detailed in the publication of PHAN VAN HIEN et al. (18).

Figure 9 demonstrates the comparitive electrophoretogram of three commercial products (erythrocupreins) mentioned in the legend and measured in different concentrations. It can be seen that only Cu-,Zn-SOD can be found in the concentrations measured.

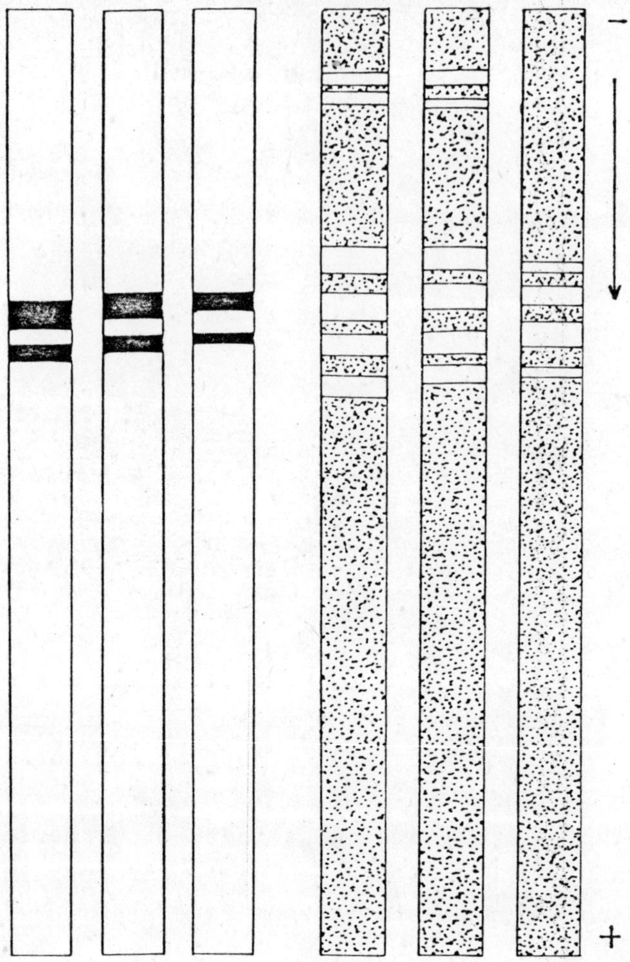

Figure 7 Polyacrylamide gel electrophoresis. Gels 1.-3. were stained for protein and from 4.-6. for SOD activity (Orgotein[R] 20, 14 and 8 μg)

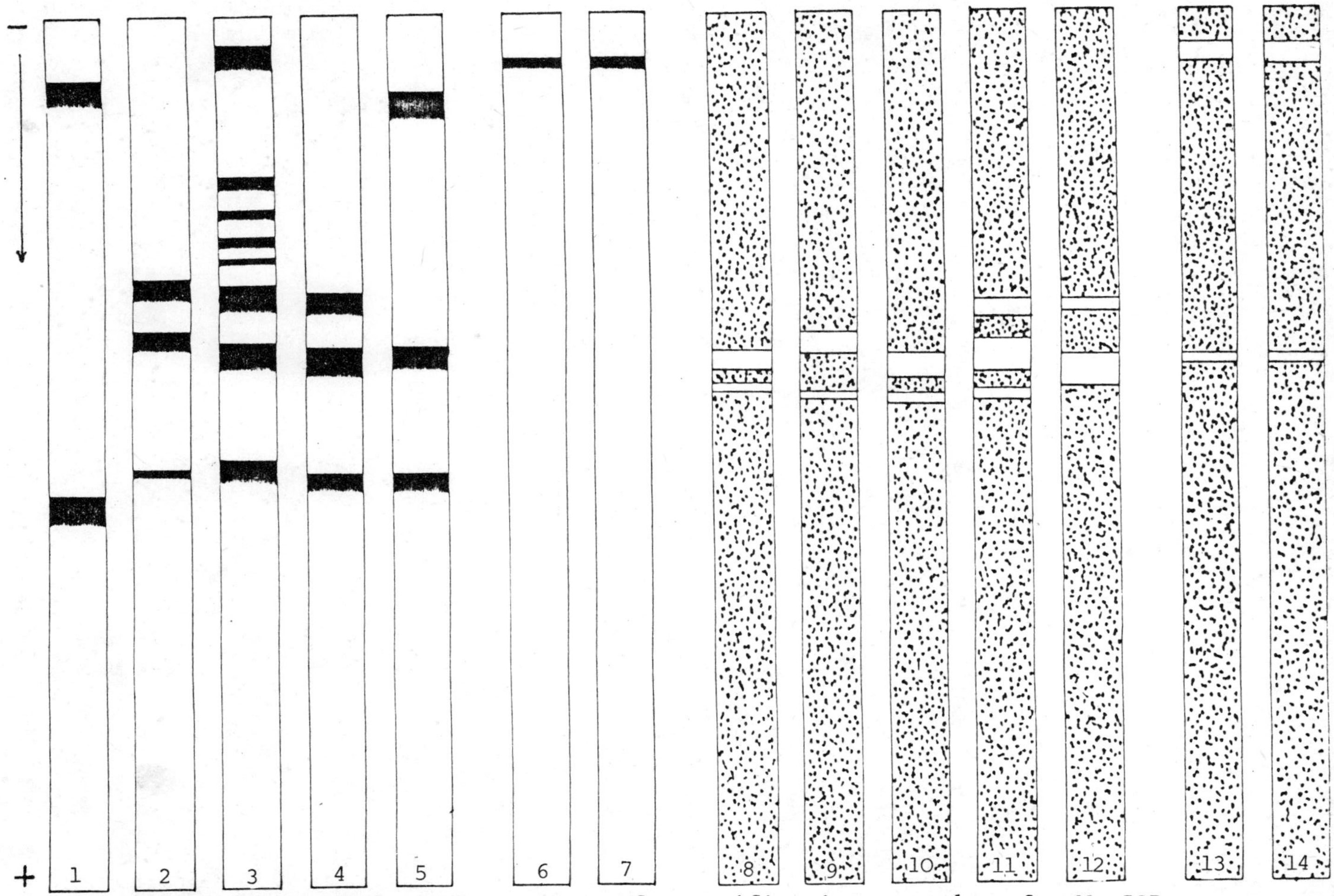

Figure 8 Electrophoresis patterns of a purification procedure for Mn-SOD from rat liver. Gels 1.-7. were stained for protein with Amido black and 8.-14. for enzyme activity with NBT

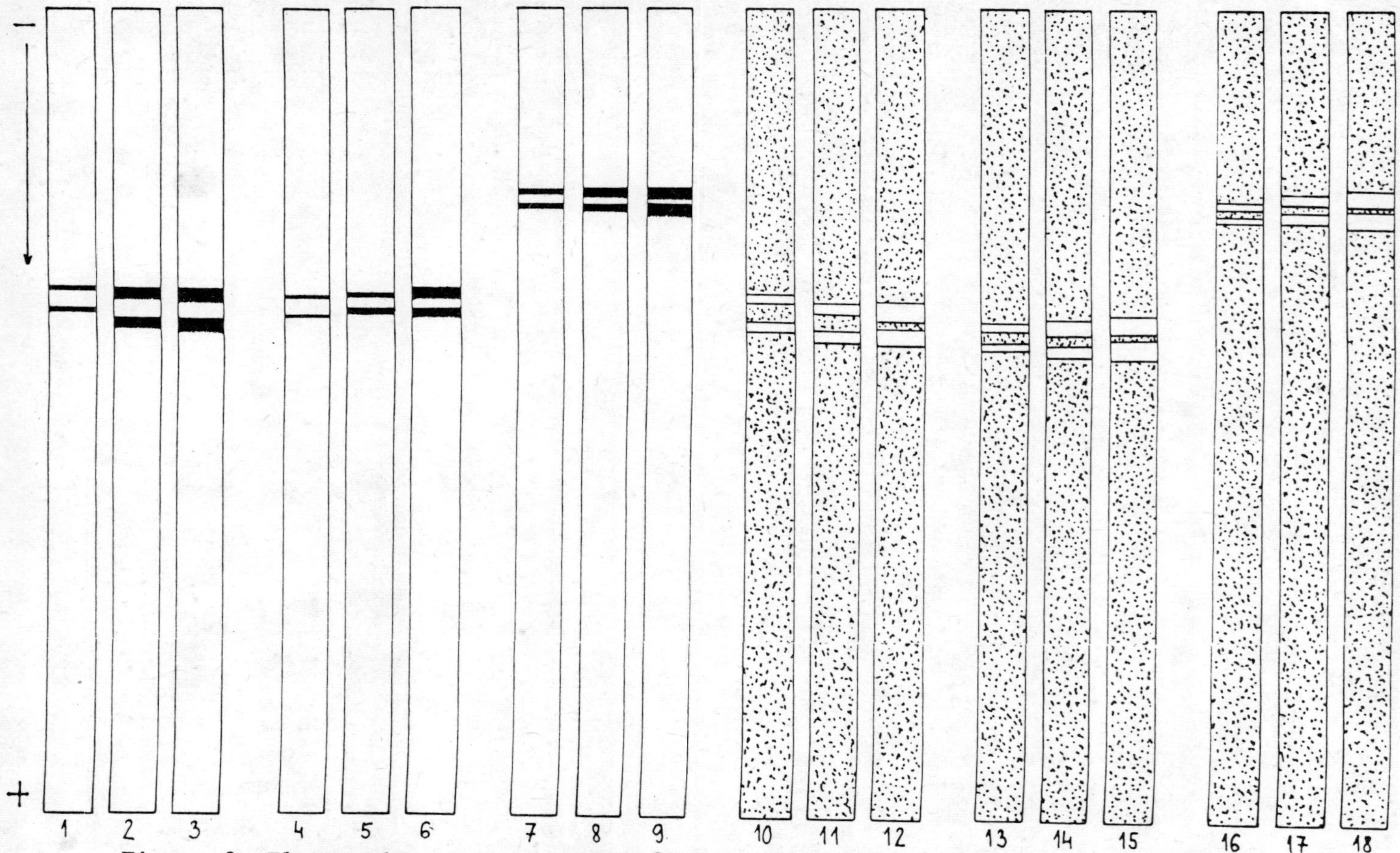

Figure 9 Electrophoretogram of different concentration of "commercial" SOD preparates from different sources. Gels 1.-3. Orgotein R 1, 5 and 10 μg, Sigma bovine RBC SOD (gels 4.-6.) and Sigma SOD from dog RBC (gels 7.-9.) stained for protein (the same amount as the Orgotein) and the following stained for enzyme activity with NBT

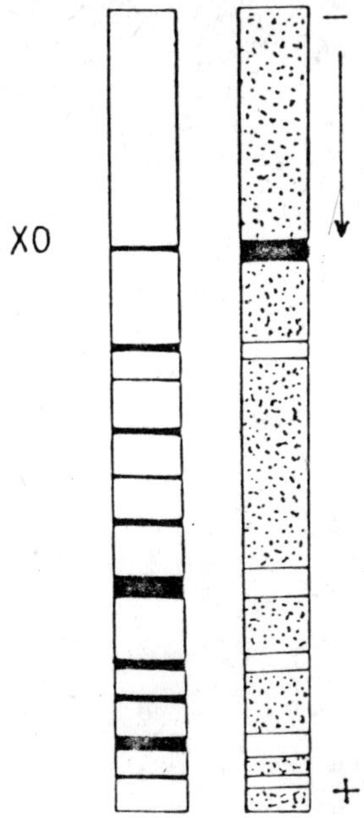

Figure 10 A typical electrophoretogram of Rhizobium meliloti, 1 column stained for protein with Coomassie Brilliant Blue and the 2nd for activity by MBT and PMS

In Fig. 10 Rhizobium meliloti xanthine oxidase (black streak) and its SODs can be seen. (In this case Fe-SOD, or rather SODs can be demonstrated beside Cu-,Zn-SOD.).

Figure 11 shows the ratio of Cu-.Zn-SOD and Mn-SOD in the organs of fish.

Our results prove that PAGE can be used well for checking the purity of the enzymes. PAGE is also suitable for separation of isoenzymes of SOD (9).

DISCUSSION

In our experiments, the Cu-,Zn-SOD originated from the human placents was purified until its activity was 550 times higher than that of the original crude preparation but using the Sephadex G-75 column, the resulting activity was increased to 2.136 higher relative than measured in the original homogenate.

In general, the activity of Cu-,Zn-SODs preparations originating from red blood cells, vegetable seeds, etc. was

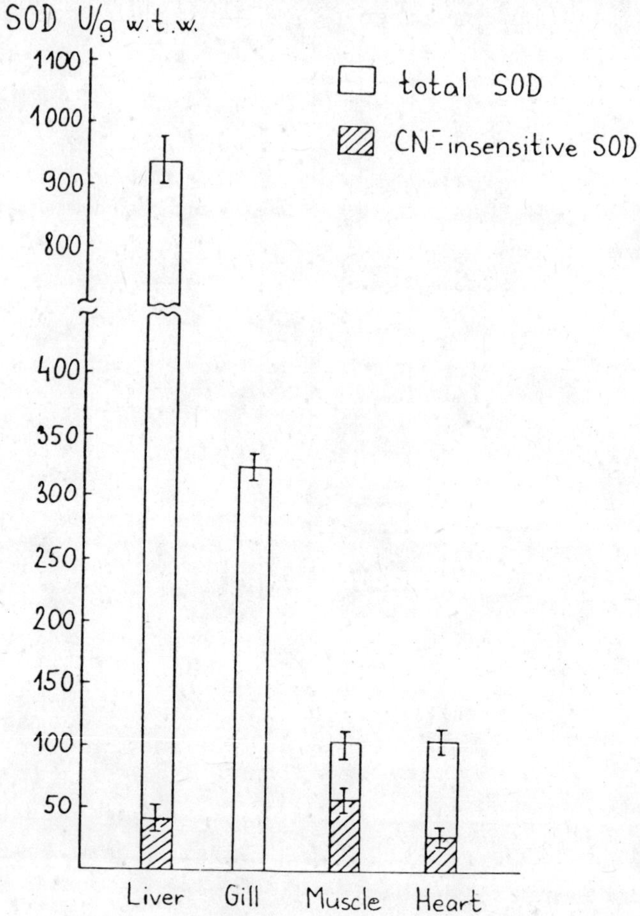

Figure 11 Total and CN^-insensitive SOD in fish organs

100-200 times higher. The Mn-SOD preparations showed about 250 times increase in purity.

The superoxide dismutases originating from Rhizobium have not been purified, they were obtained after ultrasonic digestion followed by PAGE. Our PAGE analysis resulted in two close bands and was detected by NBT whose characteristics are valid for the preparations obtained commercially, too.

The spectral characteristics and molecular size of purified SODs have also been determined. The molecular size was measured by gel filtration, standards as bovine albumin, ovalbumin, chymotrypsin and human haemoglobin were used. The molecular weight of the enzyme was finally determined by plotting the different slopes obtained against the molecular weight of standards.

SODs originating from different sources (human, animal and vegetable) were subjected to hydrolysis. The hydrolysates of the SOD preparations were determined by automatic amino acid analysis. The results obtained by the determinations of amino acids were compared with each other and with the results of the literature showing good correlation.

The striking observation is that the amino acid content of the preparations originating from subjects with diabetic disease showed oxidative demage. Similar changes of amino acids can be reached if SODs were treated with H_2O_2 solutions in vitro.

CONCLUSIONS

We have been long dealing with gel chromatographic purification, quantitative amino acid analysis and PAGE of superoxide dismutase. The investigations have practical and theoretical reasons because the purity of enzyme preparations is important from the point of the researchers. The demonstrated methods are simple, adequate enough and can be applied with good results for the purposes and aims detailed above.

(1) ASADA K., URANO M. and M. TAKAHASHI (1973) Subcellular location of superoxide dismutase in spinach leaves preparation and properties of crystalline superoxide dismutase. Eur. J. Biochem. 36, 257-266.

(2) BEAUCHAMP C. and I. FRIDOVICH (1971) Superoxide dismutase: Improved assays and an assay applicable to acrylamide gels. Anal. Biochem. 44, 276-287

(3) DAVIS B.J. (1964) Disc electrophoresis II. Methods and application to human serum proteins. Ann. N.Y. Acad. Sci. 121, 404-427

(4) FRIDOVICH I. (1975) Superoxide dismutases Ann. Rev. Biochem. 44, 147-159

(5) HALLIWELL B (1978) Biochemical mechanisms accounting for the toxic action of oxygen on living organisms(The key role of superoxide dismutase. Cell Biol. Internat. Reports. Vol. 2, No. 2. 113-128
(6) HAMILTON P.B. (196) Ion exchange chromatography of amino acids. A single column, high reso-living, fully automatic procedure. Anal. Chem. 35, 2055-64
(7) HANNING K. (1959) Erfahrungen mit der quantitativen Aminosaurebestimmung an Ionastauschersäulen und automatischer Registrierung der Ergebnisse. Clin. Chim. Acta 4, 51-57
(8) HANŪSZ B. (1974) Ph. D. Dissertation "A.J." University of Szeged
(9) HARRIS H. and HOPKINSON D.A. (1978) Handbook of enzyme electrophoresis in human genetics. Suppl. 1978. North-Holland Publ. Co. Amsterdam
(10) HARTZ, J.W. and DEUTSCH H.F. (1969) Preparation and physicochemical properties of human erythrocuprein. J. Biol. Chem. 244 4564-4572
(11) HASSAN H.M. and I. FRIDOVICH (1980) Superoxide dismutases: Detoxification of a free radical. Edited by: JAKOBY W.B. Enzymatic basis of detoxification. Vol. 1, 311-332. Academic Press, New York
(12) HUBER W. and MENANDER-HUBER K.B. (1980) Orgotein. Clinics in Rheumatic Diseases. Vol. 6 No. 3. 465-598
(13) KERESE I. and MATKOVICS B. (1973) Symp. on aminoacids. Oct. 15.-19. Brono, Czechoslovakia
(14) MARKOWICZ H., CARTWRIGHT G.E. and M.M. WINTROBE (1959) Studies on copper metabolism. XXVII. The isolation and properties of erythrocyte cuproprotein /erythrocupreinç. J. biol. Chem. 234, 40-45
(15) MATKOVICS B. (1977) The radicals of molecular oxygen and enzymatic defence against some of them. Acta Univ. Lodziensis, Seria II, 103-154
(16) MISRA H.P. and I. FRIDOVICH (1975) The role of superoxide anion in the autoxidation of epinephrine and a simple assay for superoxide dismutase. J. biol. Chem. 247, 3170-3175
(17) MOORE S. and W.H. STEIN (1948) Photometric ninhydrin method for use in the chromatography of aminoacids. J. biol. Chem. 176, 367-388
(18) PHAM VAN HIEN, KOVACS K. and B. MATKOVICS (1974) Properties of enzymes. I. Study of superoxide dismutase activity change in human placentae of different ages. Enzyme 18, 341-347
(19) SAWADA Y., OHYAMA T. and I. YAMAZAKI (1972) Preparation and physicochemical properties of green pea superoxide dismutase. Biochem. Biophys. Acta 268, 305-312

Proc. of the Symp. on Advances in Liquid Chromatogr., Szeged, Hungary, 1982

APPLICATION OF LIQUID CHROMATOGRAPHIC METHODS IN THE BIOCHEMICAL ANALYSIS OF TUMOR CELL MEMBRANES

T. KREMMER,* T. TÓTH,** L. HOLCZINGER*

*National Oncological Institute, Research Institute of Oncopathology, Department of Biochemistry, Budapest
**Eötvös Loránd University, Department of Chemical Technology, Budapest, Hungary

A wide variety of chromatographic methods are in use now contributing effectively to the preparation and analysis of complex biological materials. Cell membranes originating from normal or pathologic tissues including malignant cells represent a typical target in biomedical applications of chromatography. The characteristic features of classic and recent models of biomembranes are based on the lamellar-bilayer, fluid mosaic crystalline structure of membrane lipids in which (glyco)-protein subunits are fully or partly embedded (LENAZ, 1979). It is considered now that all these constituents determine the physico-chemical state and regulate the variable biological functions of cell membranes (BOREK et al. 1980).

Results presented here are to demonstrate the applicability of liquid chromatographic methods in the analysis of lipid and protein constituents of Ehrlich ascites tumor (EAT) cell membranes in the two different phases of tumor growth.

Abbreviations(EAT Ehrlich ascites tumor, TLC Thin layer -, GLC Gas liquid -, HPLC High performance liquid chromatography, SPH Sphingomyelin, PC, PE, PS, PI Phosphatidyl-choline, -ethanolamine, -serine, -inositol, Gly Glycolipids, MDG Mono-diglycerides, CHOL Cholesterol, FFA Free fatty acids, TG Triglycerides, CE Cholesterol esters, SDS Na-dodecyl sulfate.

MATERIALS AND METHODS

EAT cells were transplanted intraperitoneally into Swiss outbred H/Riop male mice and harvested on the 7th and 14th days of tumor growth. Tumor cell cytoplasmic membranes were prepared by the method of FORTE et al. (1973). Membrane total lipids were extracted and weighed according to FOLCH et al. (1957). Proteins were determined by the HARTREE's method (1972). Phospholipids separated by TLC were measured according to ROUSER et al. (1970). Neutral lipids were fractionated and quantified by bichromate oxidation modified by us (KREMMER et al. 1969). TLC separation of lipids was carried out on home-made and pre-coated Silica gel 60 (Merck) plates. GLC of fatty acid methyl esters was performed by capillary column technique (SP-1000, 30 m x 0.25 mm, col. oven 150 - 220°C, 3°C/min, carrier N_2 100 kPa, FID 32 x 2 . 10^{-12} AFS) on Packard 427 gas chromatograph coupled to a minigrator M-2 and linear recorder. For peak identification authentic references and in some cases selective preinjection reactions (H_2-sat) were used. Quantitative evaluation was performed by normalizing all peak areas for 100 per cent between 12:0 - 24:0. Reversed phase HPLC was carried out with Micromeritics Instrument Corp. (SY-LAB, Wien) Model 731 Column Compartment, 740 Control Module, 750 Solvent Delivery System, 752 Gradient Programmer and 786 Variable Wavelength Detector. Redestilled solvents were degassed by He. Gel chromatography was performed as described previously using an LKB Complete Column Chromatography System /KREMMER and HOLCZINGER, 1980). Membrane proteins were fractionated by polyacrylamide gel electrophoresis according to WISHER and EVANS (1975) and stained for proteins with Coomassie Brilliant Blue G-250 and for glycoproteins with periodate-Schiff reagent.

RESULTS AND DISCUSSION

TLC of membrane lipids

Membrane lipids represent a very heterogenous group of compounds containing aliphatic alcohols, amines, fatty acids, phosphate esters, carbohydrates, sterols, etc. The lipid composition of biomembranes is rather characteristic but it may differ from membrane to membrane depending on their origin (BOGGS, 1980). In general, membrane lipids are composed first of all of polar, mainly phospho- and glycolipids and cholesterol, respectively. A rather extensive fractionation of membrane polar lipids can be obtained by one-dimmensional TLC on silica gel in a solvent system of chloroform-ethylacetate-n-propanol-methanol-0.25% KCl (10:10:10:12:5). Total and fractionated phospholipids are usually measured as inorganic phosphate after oxidative digestion. Staining with phosphomolibdenic acid followed by densitometry can also be used for the in situ quantitation of polar lipids separated (Fig. 1, PETER and REYNOLDS, 1974, VASKOVSKY et al. 1975).

Membrane neutral lipids (CHOL, CE, TG, FFA, MDG) can easily be separated on silica gel in a solvent system of perolether-diethyl ether-acetic acid (85:15:2) and quantified with specific microchemical methods or by bichromate oxidation and photometry (KREMMER et al. 1969). In the last years we have used the latter technique with success in the routine analysis of membrane lipids.

Table 1 summarizes the results obtained for the lipid composition of EAT cell membranes in the two characteristic phases of tumor growth. It can be stated that there was significant fall in the membrane SPH and CHOL at the stationary phase (14 days) of tumor cell proliferation, while the TG content has doubled.

GLC of membrane fatty acids

It has been suggested that compositional changes observed in the lipids of EAT cells were in close correlation with the tumor-induced endogen mobilization of host lipids (HOLCZINGER and KREMMER, 1975, KREMMER and HOLCZINGER, 1976). In addition, characteristic differences were found between the two phases of tumor growth in the fatty acid composition of membrane lipids separated by preparative TLC. Fig 2 shows a marked in-

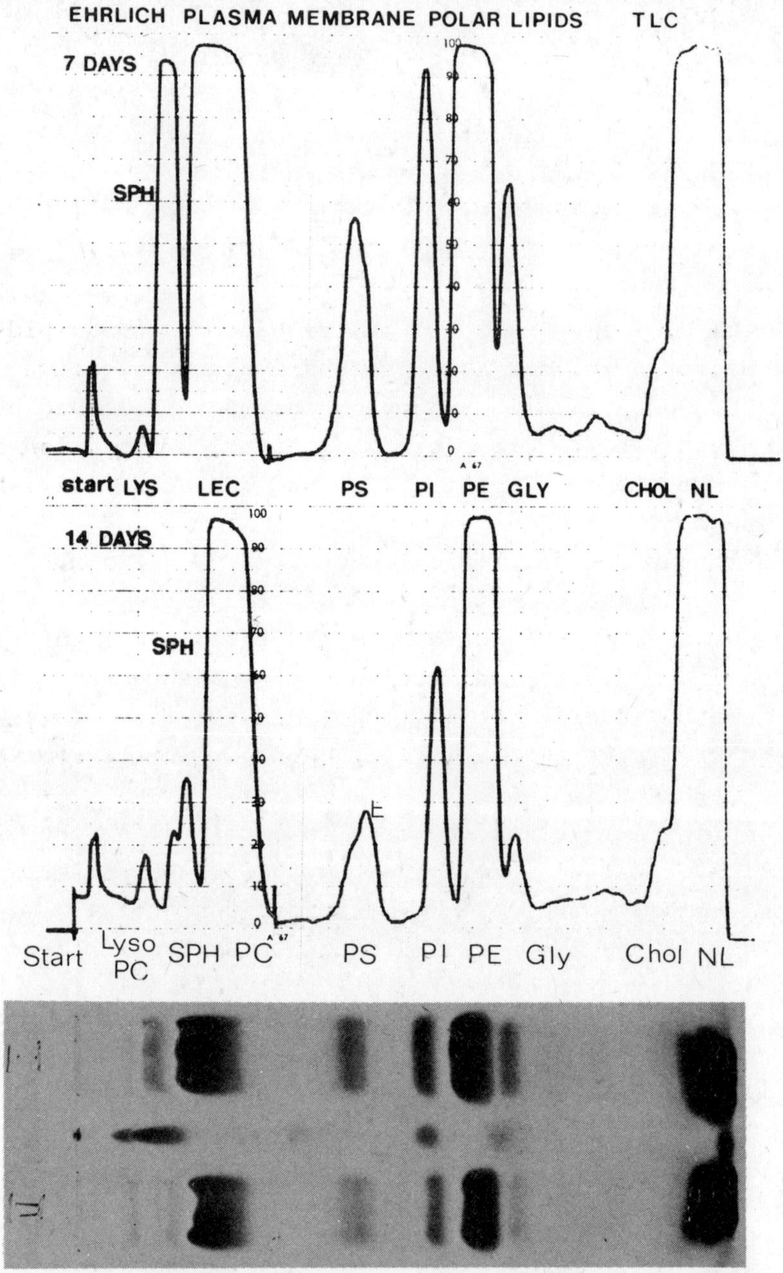

Fig. 1. TLC (see in the text) and photodensitometric scanning of EAT membrane polar lipids in the 7th (I) and 14th (II) days of tumor growth. Development: spray with 5% phosphomolibdenic acid, 10% perchloric acid in 50% ethanol and heating for 10 min at 100°C. Densitometry: Varicord Model 42B (Photovolt Corp.) at 610 nm

TABLE 1

Lipid composition of EAT cell membranes

Lipids⁺ rel.%	Ehrlich ascites tumor cell membranes 7 days	14 days
Polar lipids	67.7 ± 5.4	67.1 ± 5.2
PC	28.0 ± 0.7	28.7 ± 1.3
PE	20.6 ± 0.9	22.9 ± 2.1
SPH	7.4 ± 0.7	3.8 ± 0.7
PS	4.5 ± 0.5	5.1 ± 0.7
PI	4.5 ± 0.5	4.2 ± 0.2
LysoPC	0.4 ± 0.1	1.1 ± 0.7
Gly	1.8 ± 0.4	1.2 ± 0.5
Neutral lipids	32.3	32.9
CHOL	11.8 ± 0.9	7.7 ± 2.6
CE	4.4 ± 1.8	6.8 ± 3.0
TG	3.2 ± 1.8	7.8 ± 3.2
FFA	9.9 ± 4.3	9.1 ± 2.4
MDG	2.9 ± 1.9	1.6 ± 1.0

crease in the saturation of TG originated from the 14 day-old cell membrane (TG-II, 16:0, 17:0, 22:0) indicating an extensive incorporation of depot fatty acids into the tumor cells. In the main polar lipids (see Table 2) a chain elongation of PC fatty acids (16:0 to 18:0) was shown similar to that of neutral lipids but with a tendency of unsaturation (18:1, 18:2, 20:1). Fatty acid composition of PE represented a more pronounced tendency for unsaturation (18:2, 18:1, 20:4) with a parallel decrease of palmitic and stearic acids. In SPH a clear-cut change of 16:0 to 18:0 was detected both by GLC and HPLC. It has to be noted that such changes in the fatty acid composition of biomembranes involve basic alterations in their structure and fluidity.

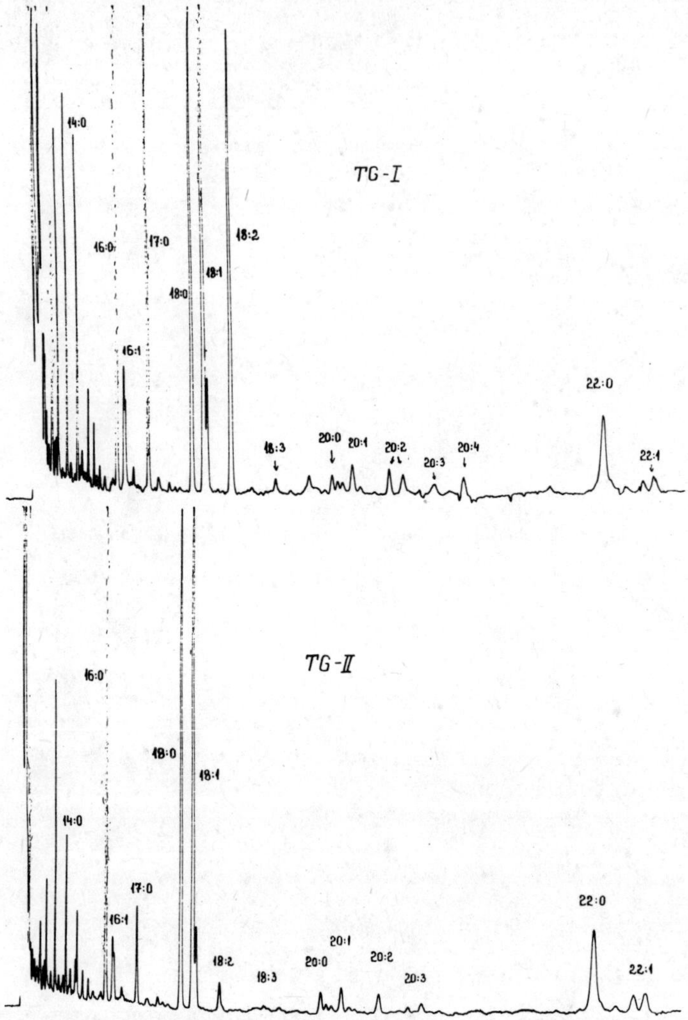

Figure 2 GLC patterns of fatty acids (methyl esters) originated from triglycerides of 7 (TG-1) and 14 (TG-II) day-old EAT cell membranes

TABLE 2

Fatty acid composition of the main polar lipid fractions originated from EAT cell membranes

Fatty acids+	PC 7+	PC 14++	PE 7	PE 14	SPH 7	SPH 14
14:0	1.0	1.0	0.5	0.5	2.2	2.2
16:1	0.8	0.9	0.6	0.8	1.0	1.8
16:0	28.2	28.3	13.6	9.8	47.0	31.2
18:2	1.6	3.4	9.5	21.7	1.0	5.0
18:1	3.2	5.8	11.0	17.7	5.0	7.2
18:0	25.9	35.7	56.4	36.9	27.1	49.1
20:4	-	-	1.3	6.2	-	-
20:3	-	-	0.3	2.0	-	-
20:2	-	-	4.0	2.1	-	-
20:1	9.7	11.8	0.6	1.0	-	-
20:0	1.1	1.5	0.9	0.6	1.0	0.5
22:1	0.9	0.8	-	-	1.6	0.5
22:0	0.4	0.3	0.4	0.3	4.3	1.5
24:1	-	-	-	-	1.7	0.5
24:0	0.7	0.6	0.4	0.4	8.1	1.5

+ Minor fatty acids are not indicated here

++ 7 - Seven, and 14 - fourteen day-old tumor cell membranes

HPLC of membrane sphingomyelins

HPLC of tumor cell SPH prepared from membrane polar lipids by TLC was performed according to JUNGALWALA et al. (1980) using bovine brain SPH (Calbiochem, Lot 400894, IR-spectra controlled) as a reference mixture. Fractionation was carried out isocratically on RP-18 column (HP 25x0.46 cm, 5μ) with methanol-5 mM phosphate buffer, pH 7.4 (95:5 v/v) solvent system, flow rate: 1.5 ml/mn (P_{max}:450 kPsi) at ambient temperature. Detection was at 205 nm, SPH samples (300-500 μg) were injected in 20-40 μl volumes. Individual SPH peaks were collected and analysed for fatty acids by GLC,

In our work the HPLC pattern of bovine brain SPH was identical to that reported by JUNGALWALA et al. (1980). In comparison, SPH profiles obtained for tumor cell membranes showed two differences:1) they had a pattern characteristic of EAT, 2) a relative increase in the stearic acid-rich molecular species of membrane SPH was demonstrated at the stationary phase (14 days) of tumor cell proliferation (Fig. 3). This coincided with the GLC data taken for the whole membrane SPH (Table 2) as well as for the individual HPLC peaks collected.

Considering the significance of SPH in biological membranes and their functional similarity to that of cholesterol (BARENHOLZ and THOMPSON, 1980) it is suggested that the fall of SPH and CHOL in EAT cell membranes during the tumor growth (Table 1)

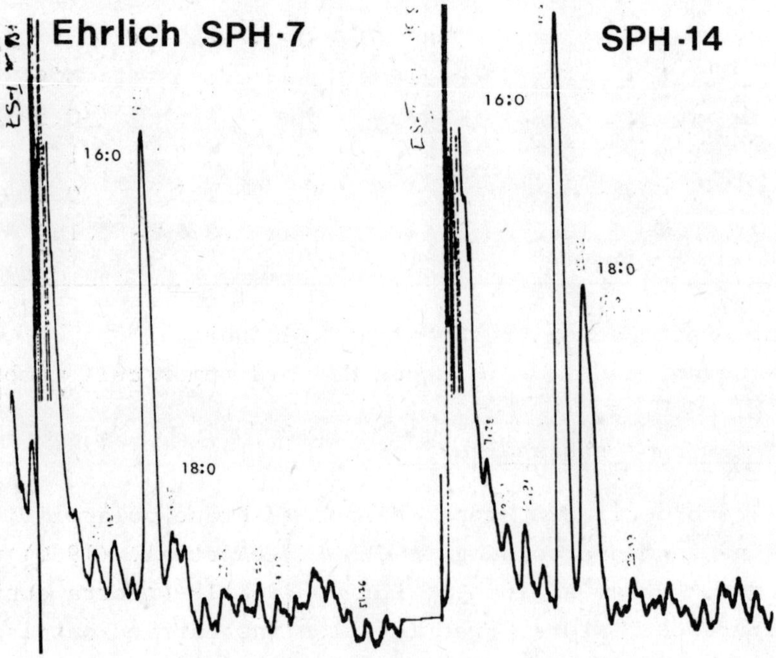

Fig.3 HPLC fractionation of SPH originated from the 7 and 14 days old EAT cell membranes

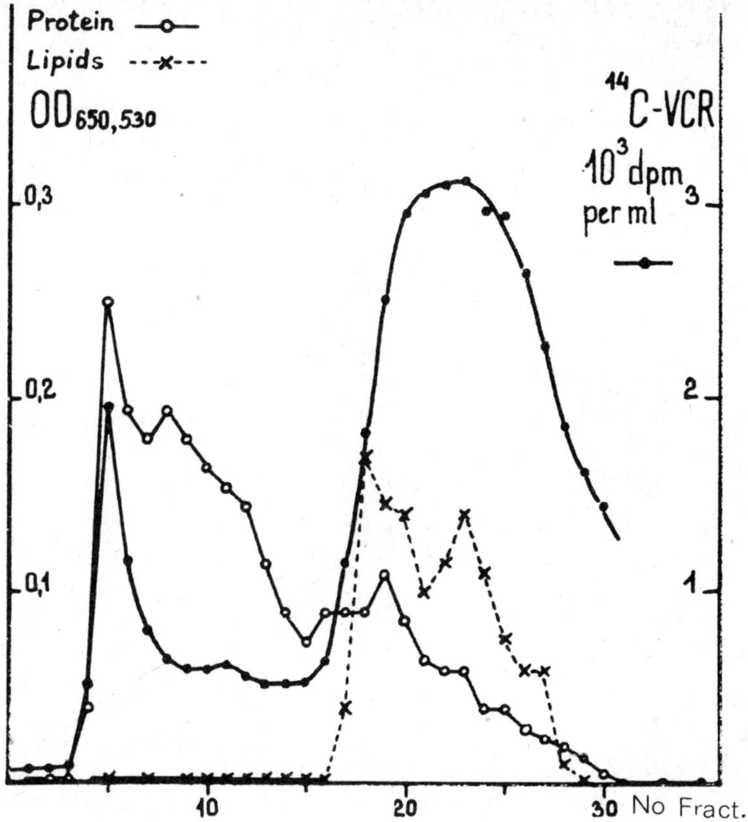

Fig. 4 Detergent gel chromatography of EAT cell membranes on Sepharose 4B column (90x2.54 cm, V_t:450 ml) eluted with 2 mM $HaHCO_3$ (pH 7.5) containing 1% Triton X-100

has a special importance in relation to the membrane structure and function.

Gel chromatography of membrane proteins

Membrane proteins rich in carbohydrates and related compounds represent the hightly insoluble structural and functional elements of biomembranes. For their preparation and characterization it is necessary to solubilize them by detergents to get well preserved and enzymatically active (glyco)protein subunits and solubilized lipid micelles. A general method separating

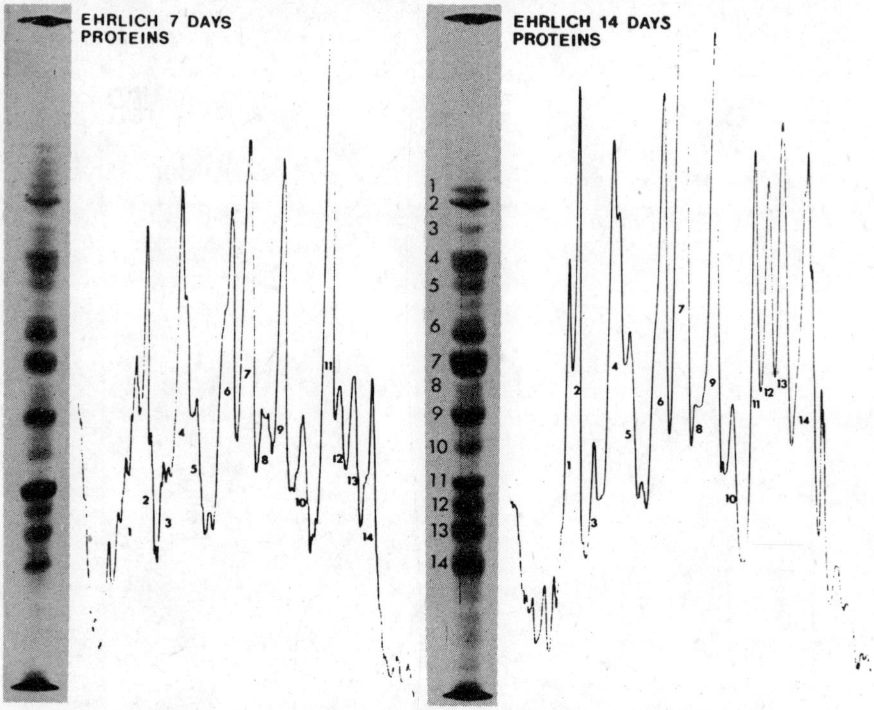

Fig. 5 Gel electropherograms and photodensitometric scans of EAR cell membrane proteins in the 7th and 14th days of tumor growth. Labor MIM OE-503 Densitometer (590 nm) coupled to MTA Kutesz Typ. 185 Recorder

membrane protein subunits according to their virtual molecular weight is column chromatography using gel particles of appropriate pore size. This principle proved to be useful for the study of drug-cell membrane interactions too (KREMMER and HOLCZINGER, 1980). Detergent gel chromatography of ^{14}C-Vincristine-pretreated EAT cell membrane solubilized with Triton X-100 (Fig. 4) indicated that the major part of membrane-bound drug was associated with the lipid micelles and a small proportion only was bound to the protein subunits.

Electrochromatography of membrane proteins

Electrochromatographic (electrophoretic) procedures permit detailed fractionation of solubilized membrane proteins according to their charge and size. High resolution can be obtained using

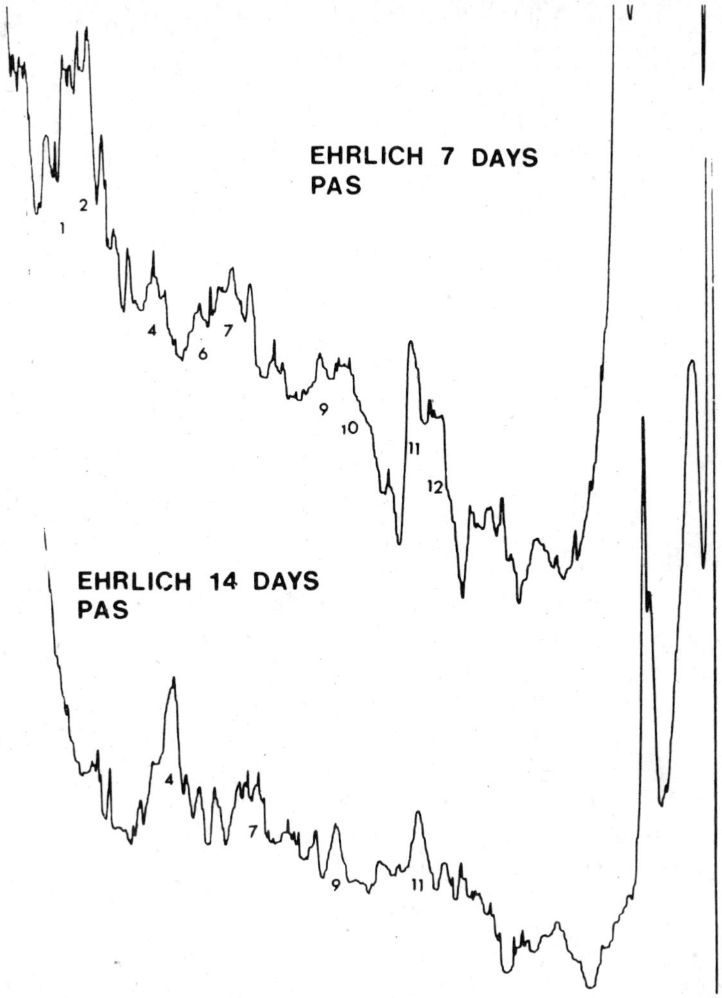

Fig. 6 Photodensitometric profiles of the 7 and 14 day-old EAT cell membrane glycoproteins. Electropherograms of Fig. 5 were stained for carbohydrates and scanned at 560 nm

polyacrylamide gels where gel filtration is combined with the charge effect in the process of separation. In practice, horizontal and vertical, disc or slab techniques are used. Fig. 5 shows the results of the vertical-disc electrophoretic fractionation of SDS-solubilized EAT cell membrane proteins in the 7th and 14th days of tumor growth. Marked differences in the (glyco)protein subunit patterns between the two phases of tumor growth can be quantified by photodensitometric scanning of electropherograms stained either for proteins (Fig. 5) or for carbohydrates (Fig. 6).

Summarizing our results it can be stated that application of combined chromatographic methods in the analysis of tumor cell membrane constituents gives new possibilities to detect subtle molecular alterations occurring during the tumor growth contributing to the better understanding of structure and function of biological membranes.

Acknowledgement

We gratefully acknowledge the Micrometrics Instrument Corp. and Mr. G. FRANZL (SY-LAB, Wien) for the generous support and Mrs. VERA SCHLEURING and Mrs. ANDREA PENZES for the skilful assistance.

REFERENCES

(1) BARENHOLZ, Y. and THOMPSON, T.E. (1980) Biochim. Biophys. Acta 604, 129.
(2) BOGGS, J.M. (1980) Can. J. Biochem. 58 755.
(3) BOREK, C., FENOGLIO, C.M. and KING, D.W. (1980) Aging, Cancer and Cell Membranes, Part II, G. Thieme Verlag, Stuttgart-New York, p. 250.
(4) FOLCH, J.G., LEES, M. and SLOANE-STANLEY, G.H. (1957) J. Biol. Chem. 226, 497.
(5) FORTE, J.G., FORTE, T.M. and HEINZ, E. (1973) Biochim. Biophys. Acta 298, 827.
(6) HARTREE, E.F. (1972) Anal. Biochem. 48, 422.
(7) HOLCZINGER, L. and KREMMER, T. (1975) Acta Histochemica Suppl. Bd. XV. 35.
(8) JUNGALWALA, F.B., HAYSSEN, V., PASQUINI, J.M. and McCLUER, R.H. (1979) J. Lipid Res. 20, 579.
(9) KREMMER, T., FERENCZY, E. and POSCH, E. (1969) Chromatographia 2, 142.
(10) KREMMER, T. and HOLCZINGER, L. (1976) Acta Morphologica Acad. Sci. Hung. 24, 369.
(11) KREMMER, T. and HOLCZINGER, L. (1980) J. Chromatogr. 191, 287.
(12) LENAZ, G. (1979) The role of lipids in the structure and function of membranes. In Subcellular Biochemistry, Vol. 6. p. 233. Ed. Roldyn, B.D., Plenum Press, New York.

(13) PETER, F. and REYNOLDS, R.G. (1974) Clin. Biochem. 7, 371.
(14) ROUSER, G., FLEISCHER, S. and YAMAMOTO, A. (1970) Lipids 5, 494.
(15) VASKOVSKY, V.E., KOSTETSKY, E.Y. and VASENDIN, I.M. (1975) J. Chromatogr. 114, 129.
(16) WISHER, M. H. and EVANS, W.H. (1975) Biochem. J. 146, 375.

Proc. of the Symp. on Advances in Liquid Chromatogr., Szeged, Hungary, 1982

NOVEL PURIFICATION PROCEDURE OF ALPHA-FETOPROTEIN

I. MOLNÁR

Human Serobacteriological Research Institute, Gödöllő, Hungary

SUMMARY

Human alpha-fetoprotein has been isolated from fetus by a four-step procedure sequence using salting-out with ammonium sulfate, Blue-Sepharose affinity chromatography, concanavalin A-affinity chromatography and chromatofocusing. Purified alpha-fetoprotein was used for immunisation in order to yield antisera against human alpha-fetoprotein. Antibody produced by immunisation was applied to prepare an appropriate immunoadsorbent for immuno-affinity chromatography.

INTRODUCTION

Alpha-fetoprotein is a fetal serum glycoprotein present in vertebrates. Usually it is found only in very small amounts in human adult serum, but it is well established that the occurrence of alpha-fetoprotein in blood serum is associated both with ontogenesis and with primary liver cancer in several species including man (1). A significant increased concentration of alpha-fetoprotein in amniotic fluid is associated with open neural tube lesions in the developing fetus (2). The clinical significance of alpha-fetoprotein as a marker in the diagnosis and monitoring of certain types of cancer and noncancerous hepatitic diseases as well as screening for neural tube defects have been extensively published (3-6).

Purification of alpha-fetoprotein by salting-out, ion-exchange chromatography, gel filtration, affinity chromatography and immunochemical methods have been reviewed by Twomey et al. (7), Wu et al. (8), Pagé (9), Ruoslahti (10), Lai et al.(11), Young et al. (12,13) and Kapadia et al. (14). Accordingly, a four-step purification procedure is presented here employing a combination of physicochemical and immunochemical approaches to prepare alpha-fetoprotein that was free of albumin and other serum proteins.

MATERIAL AND METHODS

Materials

Concanavalin A-Sepharose 4B, Blue-Sepharose CL-6B, Sepharose 4B were all purchased from Pharmacia Fine Chemicals /Uppsala, Sweden/. Servacel DEAE 52, 1-O-methyl- α -D-glucopyranoside, phenylmethylsulphonyl-fluoride were purchased from Serva Fine Biochemicals /Heidelberg,GFR/. Cyanogen bromide was from Fluka /Buchs,Switzerland/ and antiserum to human alpha-fetoprotein, albumin, transferrin and other antiseras were obtained from Dakopatts. Carrier ampholyte was home-made synthetised according to Righetti (15).

Purification procedure

For the preparation of alpha-fetoprotein a fetal material kept at - 20°C was used. 500 g of fetal material were defrozen at + 4 °C and then homogenized in 1250 ml of Tris/HCl buffer, 0.05 M, pH 7.4, containing 0.01 M EDTA, 0.02 % sodium azide and 0.0015 M phenylmethylsulphonyl-fluoride. The homogenized material was then put in a Becker and kept under agitation at + 4°C for 18 hours. It was subsequently centrifuged at 3000 r.p.m. for 30 minutes at + 4°C. The supernatant liquid was separated from the solid that was discarded and raised to pH 6.0 by means of HCL solution. After measuring the volume it was treated, under agitation, by means of solid ammonium sulfate at the rate of 286.7 g per litre of solution. When the salt was passed into the solution the precipitate formed was separated by centrifugation and discarded. 205 g of ammonium sulfate per litre of solution were added to the supernatant. Once the dissolution of the salt was completed it was centrifuged at 5500 cycles for 30 minutes at + 4°C. The supernatant was then discarded and the residual material was collected and vacuum dried material served as starting material for further purification.

Blue-Sepharose affinitiy chromatography

The starting material was applied to a 1.6 x 14 cm column containing Blue-Sepharose CL-6B. The partially purified material was dissolved and dialysed against 50 mM Tris/HCl buffer, pH 7.5 which contained 100 mM KCl. The equilibrium buffer was also used to elute the proteins by descending flow which did not bind to Blue-Sepharose. The albumin bound to the column was eluted with 50 mM Tris/HCl buffer, pH 7.5, containing 2.0 M KCl. The elution profile was monitored by measuring the absorbance at 280 nm. The alpha-fetoprotein containing fractions were pooled and dialysed against 0.1 M acetate buffer, pH 6.0, containing 1.0 M NaCl, 1 mM $CaCl_2$, 1 mM $MnCl_2$, and 1 mM $MgCl_2$. The solution was concentrated to a volume of 8 ml by means of a PM-10 membrane.

Concanavalin A-Sepharose affinity chromatography

The concentrate was applied to a 1.6 x 25 cm column filled with concanavalin A-Sepharose. Prior to use, the column was equilibrated with 0.1 M acetate buffer, pH 6.0, containing 1.0 M NaCl, 1 mM $CaCl_2$, 1 mM $MnCl_2$, and 1 mM $MgCl_2$. The sample was applied at room temperature and eluted at a rate of 20 ml per hour. The elution was recorded by monitoring absorbance at 280 nm with an Isco UA-5 absorbance monitor. Fractions of 2 ml were collected. Protein that did not bind to concanavalin A was eluted with the equilibrium buffer. After the first peak of protein was recorded, the eluting medium was changed to an equilibrium buffer containing 2 % 1-O-methyl- α-D-glucopyranoside.

Chromatofocusing

The pooled fractions eluted from concanavalin A were concentrated by ultrafiltration with PM-10 membrane and the protein samples were applied to chromatofocusing after extensive dialysis against the start buffer. For the experiments a column of 1 x 10 cm of DEAE cellulose was initially equilibrated in 10 mM phosphate buffer, pH 7.5. Nonadsorbed proteins were eluted with this starting buffer and the adsorbed protein were eluted in succession with

/I/ 104 ml of pH 4-6 ampholyte, pH 5.5
/II/ 80 ml of pH 3.5 - 5 ampholyte, pH 4.65
/III/ 0.3 M NaCl in the starting buffer

Elution was carried out at a flow rate of 15 ml per hour and fractions of 2.4 ml were collected. The effluents were monitored for absorbance at 280 nm with an Isco UA-5 adsorbance monitor. The ampholyte used was home synthetised and fractionated to narrow pH ranges.

Antisera

Goat anti-human alpha-fetoprotein was produced by immunizing the animal with 380 µg of alpha-fetoprotein in complete Freund's adjuvant at multiple intradermal sites. The injections were given three times at 14 day intervals and the animal was bled after the last injection.

Immunoadsorbent

Immunoglobulin fraction was prepared from antisera by precipitation with sodium sulfate / 18 g of solid sodium-sulfate/ 100 ml of serum/. The precipitate was dissolved in distilled water, dialysed against water and liophylised after removal of the precipitate formed during dialysis. The freeze-dried material was dissolved in 0.1 M $NaHCO_3$, pH 9.7, at a concentrate of 25 mg/ml and was coupled to activated Sepharose 4B. The Sepharose 4B was activated with cyanogen bromide using the method of Porath (16). The immunoadsorbent was packed in a 1.6 x 20 cm column and washed with 50 mM phosphate buffer, pH 7.5, containing 0.5 M NaCl. The column was equilibrated prior to use with this buffer. Crude material containing alpha-fetoprotein was applied to the immunosorbent column and was eluted with the equilibrium buffer to wash out the unbound proteins. In succession the bound proteins were then eluted with 0.35 M glycine/HCl buffer pH 2.4, containing 0.1 M glucose and finally with 8 M urea. Descending chromatography was carried out with a flow rate of 6 ml per hour. Fractions of 1.2 ml were collected and the elution profile was monitored by measuring the absorbance at 280 nm with an Isco UA-5 absorbance monitor.

Analytical procedures

Quantitation of alpha-fetoprotein and albumin was performed employing the radial immunodiffusion technique of Mancini et al. (17). The purity of the chromatographic fractions was studied by immunoelectrophoresis according to Grabar and Williams (18) and crossed immunoelectrophoresis according to Laurell (19).

Ouchterlony double-diffusion analysis was carried out as was described by Ouchterlony (20).

RESULTS AND DISCUSSION

Blue-Sepharose affinity chromatography was performed essentially to remove albumin selectively from the preparation. All of the alpha-fetoproteins applied to the Blue-Sepharose were recorded in the unbound fraction eluted with the equilibrium buffer, but with these buffering conditions only 65 per cent of albumin was removed from the preparation containing alpha-fetoprotein.

Figure 1 illustrates the elution profile obtained from concanavalin A affinity chromatography of preparations obtained from Blue-Sepharose. Two major absorbance peaks were observed. The first peak represents proteins not bound to the column and to be eluted with the equilibrium buffer, the second appeared after the change of the eluting medium to the equilibrium buffer containing 2 % of 1-O-methyl-α - D-glucopyranoside. Immunoelectrophoresis assessment revealed that the fractions represented by the first peak consisted of albumin and reacted with anti-albumin but not with anti-alpha-fetoprotein. Proteins in the pooled fractions of the second peak had electrophoretic mobilities corresponding to those of alpha-fetoprotein and beside these three other proteins including albumin. Ouchterlony double-diffusion analysis confirmed the presence of alpha-fetoprotein antigen. However, a reaction with antiserum to albumin demonstrated that some albumin was also present.

Several previous reports (9,21) definitely stated that chromatography on concanavalin A would completely separate the alpha-fetoprotein from albumin, but this presumably is a matter of the sensitivity of the detection method. Twomay et al.(7) was of a different opinion.

Figure 2 demonstrates the separation of alpha-fetoprotein by chromatofocusing technique. Four major absorbance peaks were observed and after screening the fractions for albumin and alpha-fetoprotein by double-immunodiffusion and testing them

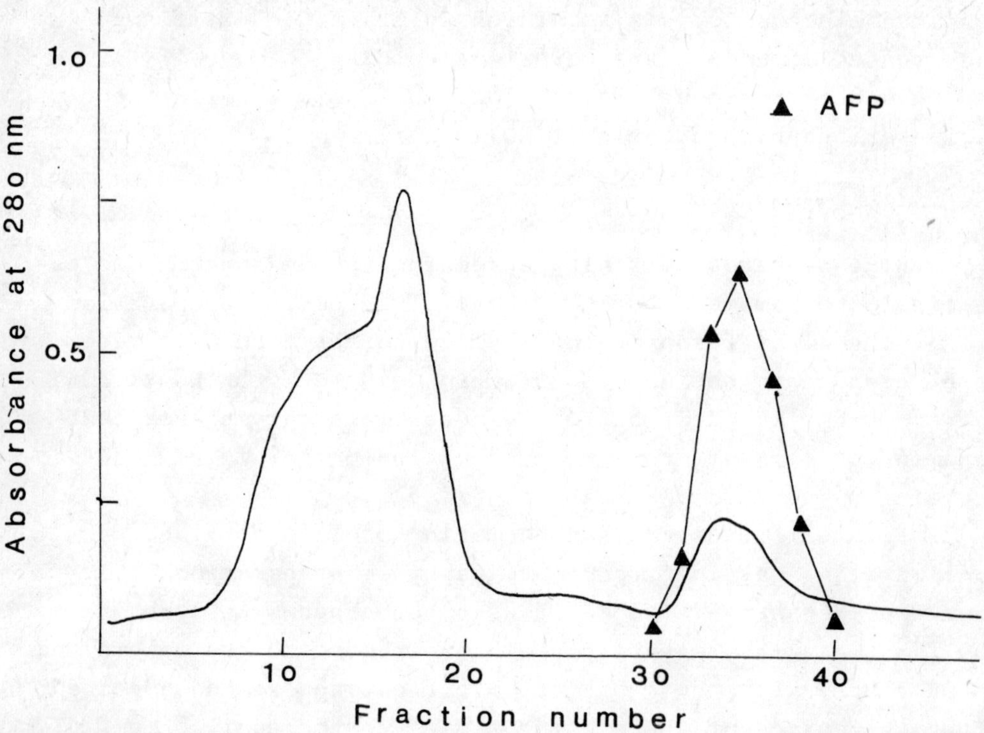

Figure 1. Concanavalin A-Sepharose 4B column chromatography of fetal preparations Starting buffer was 55 ml of 0.1 M acetate buffer, pH 6.0, containing 1.0 M NaCl, 1 mM $CaCl_2$, 1 mM $MnCl_2$, and 1 mM $MgCl_2$, followed by 125 ml of the same starting buffer, containing 2 % 1-O-methyl-α -D-glucopyranoside Absorbance at 280 nm /——/ and alpha-feto-protein /—▲—▲—/

with immunoelectrophoresis it was revealed that the third peak consisted of albumin and the fourth contained the alpha-feto-protein without albumin and other proteins as contaminants. An internally generated pH gradient was established due to the buffering capacity of ampholyte as it was shown on the figure. The concentration of ampholyte used was below that which was normally used and did not cause elution of proteins if the pH was above their isoelectric points. It can be seen from the figure that a smooth pH gradient was formed and the proteins were eluted as defined and sharp peaks. When 0.3 M NaCl was applied immediately after the third peak had emerged from the

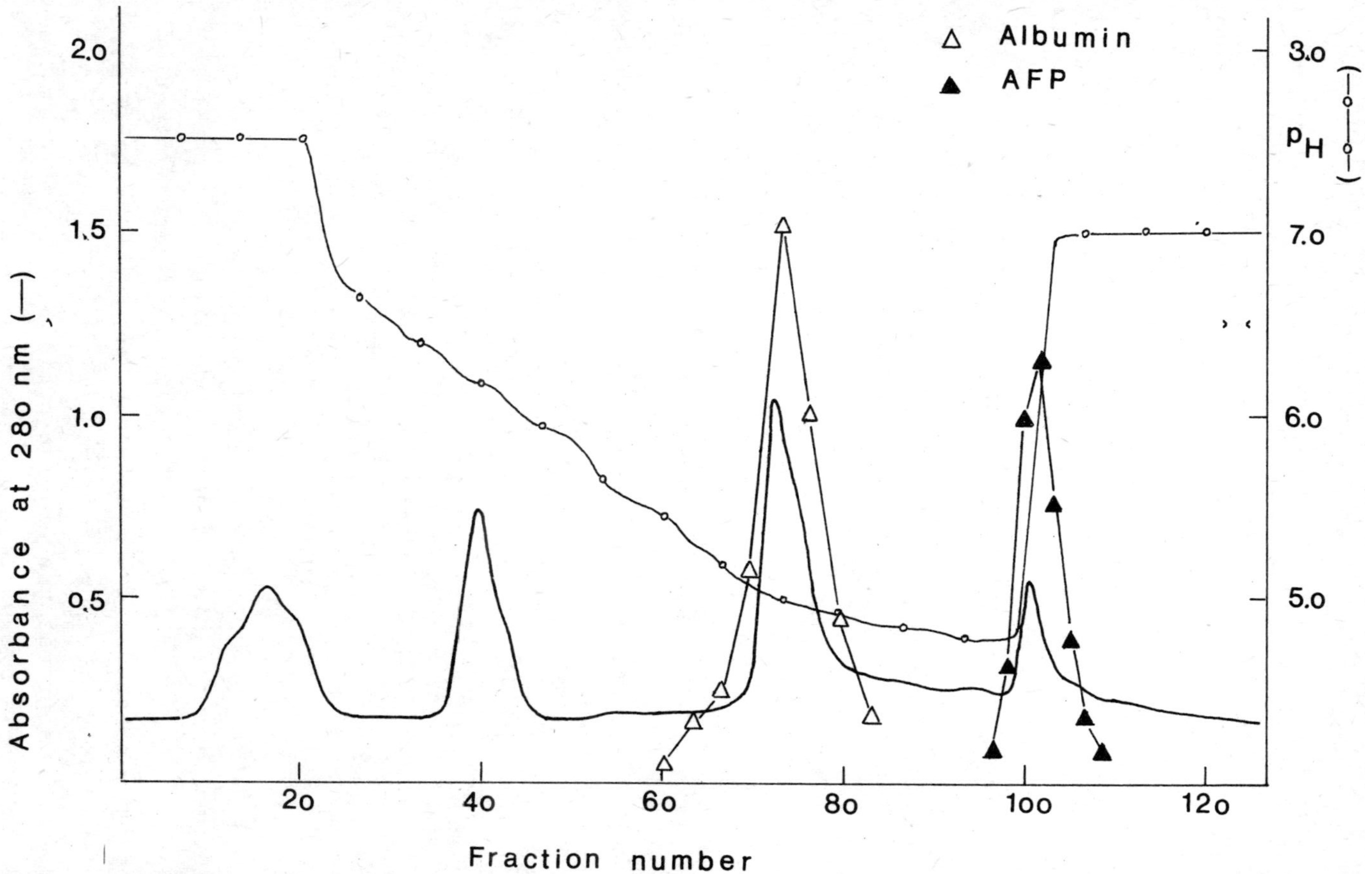

Figure 2. Chromatofocusing for the separation of alpha-fetoprotein. The column, initially equilibrated with 10 mM phosphate buffer, pH 7.5, was eluted in succession with /I/ 104 ml of pH 4-6 ampholyte, pH 5.5, /II/ 80 ml of pH 3.5-5 ampholyte, pH 4.65, /III/ 0.3 M NaCl in the starting buffer. The thick and thin lines indicate absorbance at 280 nm and pH gradient, respectively. Albumin /–△–△–△–/ and alpha-fetoprotein /–▲–▲–▲–/

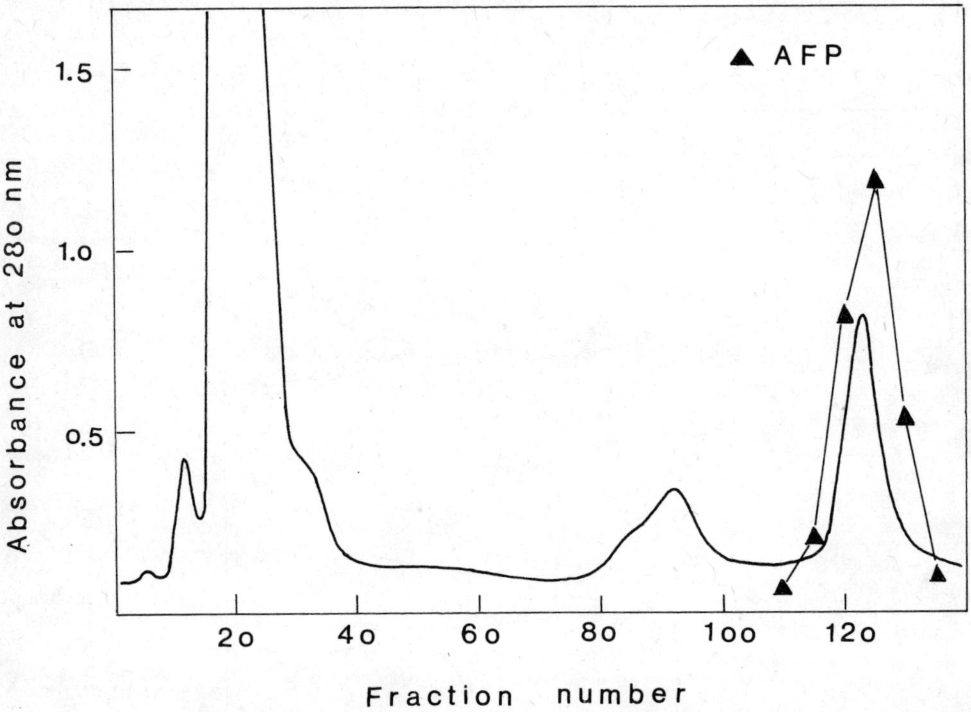

Figure 3. Immunoaffinity chromatography of alpha-fetoprotein containing material on a column of Sepharose-conjugated γ-globulin from anti-human alpha-fetoprotein serum. The column was washed prior to use with 50 mM phosphate buffer. pH 7.5, containing 0.5 M NaCl. The non-bound proteins were eluted with the same buffer, followed by 0.35 M glycine/HCl buffer, pH 2.4, containing 0.1 M glucose, and finally by elution with 8 M urea. Absorbance at 280 nm /——/ and alpha-fetoprotein /▲–▲ – ▲ –/

column, alpha-fetoprotein eluted as the last peak.

During the purification process 40 per cent of alpha-fetoprotein was lost but this process can be applied for eliminating the serum proteins especially albumin combined with alpha-fetoprotein which renders its purification difficult.

Figure 3 illustrates the elution profile obtained from the immunoaffinity chromatography of crude material yielded with salting-out, on the anti-human alpha-fetoprotein immunoadsorbent column. The first peak represents proteins not bound to the immunoadsorbent. Bound proteins were eluted in the second peak with glycine/HCl buffer as eluent, the fractions of the second peak did not contain any alpha-fetoprotein as was confirmed by Ouchterlony double-diffusion. The third peak eluted with 8 M urea revealed that alpha-fetoprotein was present. The purified alpha-fetoprotein obtained by immunoaffinity chromatography was assessed by double-diffusion and immunoelectrophoresis and was found to be contaminated with albumin and was further purified with chromatofocusing as was described previously.

In conclusion, pure human alpha-fetoprotein can be isolated in appropriate yields using appropriately selected immunochemical and physicochemical methods.

REFERENCES

(1) ABELEV,G.I., Adv.Cancer Res. 14 259 /1971/

(2) MASSEYEFF,R., Pathol.Biol. 20 703 /1972/

(3) ISHIGURO,T.,MATSUKURA,S. and MURANAKA,H., Jap.J.Clin.Pathol. 19 157 /1971/

(4) WEPSIC,H.T. and SELL,S., Prog.Exp.Tumor Res. 19 297 /1974/

(5) BROCK,D.J.H., Child's Brain 2 1 /1976/

(6) LAU,H.L. and LINKINS,S.E., Am.J.Obstet.Gynecol.124 533 /1976/

(7) TWOMEY,S.L. and SWEET,R.V., Clin.Chem. 22 1306 /1976/

(8) WU,J.T.,MONIR-VAGHEFI,S.M. and WATERHOUSE,W.J., Biochem. Med. 24 210 /1980/

(9) PAGE,M., Can.J.Biochem. 51 1213 /1973/

(10) RUOSLAHTI,E., J.Immunol. 121 1687 /1978/

(11) LAI,P.C.W., PETERS,E.H.,LORSCHEIDER,F.L., Biochem.Biophys. Acta 493 201 /1977/

(12) YOUNG,J.L. and WEBB,B.A., Science Tools, 25 54 /1978/

(13) YOUNG,J.L., REID,R.G. and CRAWFORD,J.W., Clin.Chim.Acta 69 11 /1976/

(14) KAPADIA,G.G.,KORTRIGHT,K.H.,LEE,S.Y., Prep.Biochem. 9 109 /1979/

(15) RIGHETTI,P.G., PAGANI,M. and GIANAZZA,E.J., J.Chromatogr. 109 341 /1975/

(16) PORATH,J., AXEN,R. and ERNBACK,S., Nature /London/, 215 1491 /1967/

(17) MANCINI,G.,CARBONARA,A.O. and HEREMANS,J.F., Immunochemistry 2 235 /1965/

(18) GRABAR,P.,WILLIAMS,C.A., Biochim.Biophys.Acta 10 193 /1953/

(19) LAURELL,C.B., Anal.Biochem. 10 358 /1965/

(20) OUCHTERLONY,O., Arkiv.Kemi 1 43 /1949/

(21) SMITH,C.J. and KELLEHER,P.C., Biochem.Biophys.Acta 317 231 /1973/

LIST OF CONTRIBUTORS

AIROLDI, L.
Department of Dermatology, Yale University, Medical School, New Haven, Connecticut, USA.

BALÁSPIRI, L.
Department of Medical Chemistry, University of Medicine, Szeged, Hungary.

BELENKII, B.G.
Institute of High Molecular Compounds, USSR Academy of Sciences, Leningrad, USSR.

BOJARSKI, J.
Department of Organic Chemistry, N. Copernicus Academy of Medicine, Cracow, Poland.

BUDVÁRI-BÁRÁNY, ZS.
Institute of Pharmaceutical Chemistry, Semmelweis University of Medicine, Budapest, Hungary.

ČERVENKA, J.
Department of Pediatrics, Charles University, Prague, Czechoslovakia.

DAVILA-HUERTA, G.
Department of Dermatology, Yale University, Medical School, New Haven, Connecticut, USA.

DEYL, Z.
Institute of Physiology, Czechoslovak Academy of Sciences, Prague, Czechoslovakia.

EKIERT, L.
Department of Organic Chemistry, N. Copernicus Academy of Medicine, Cracow, Poland.

GANKINA, I.S.
Institute of High Molecular Compounds, USSR Academy of Sciences, Leningrad, USSR.

GLÖCKNER, G.
Department of Chemistry, Technical University of Dresden, Dresden, German Democratic Republic.

HANKÓ-NOVÁK, K.
Institute of Pharmaceutical Chemistry, Semmelweis University of Medicine, Budapest, Hungary

HERMECZ, I.
Chinoin Pharmaceutical Works, Budapest, Hungary.

HOLCZINGER, L.
Department of Biochemistry, Research Institute of Oncopathology, National Oncological Institute, Budapest, Hungary.

HOLLÓSI, I.
National Institute of Sport-Medicine, Budapest, Hungary.

HORVÁTH, Cs.
Department of Chemical Engineering, Yale University, New Haven, Connecticut, USA.

HYANEK, J.
Department of Pediatrics, Charles University, Prague, Czechoslovakia.

ISSAQ, H.J.
Frederick Cancer Research Facility, National Cancer Institute, Frederick, Maryland, USA.

JANÁKY, T.
University of Medicine, Szeged, Hungary.

KALÁSZ, H.
Department of Pharmacology, Semmelweis University of Medicine, Budapest, Hungary.

KERECSEN, L.
Department of Pharmacology, Semmelweis University of Medicine, Budapest, Hungary.

KISS, P.
Department of Internal Medicine, Apáthy Hospital of Pediatrics, Budapest, Hungary.

KOLCOVA, V.
Pediatric Research Institute, Brno, Czechoslovakia.

KOSTIUK, I.O.
Institute of High Molecular Compounds, USSR Academy of Sciences, Leningrad, USSR.

KOVÁCS, J.
Section of Pediatrics, Péterfy Hospital, Budapest, Hungary.

KREMMER, T.
Department of Biochemistry, Research Institute of Oncopathology, National Oncological Institute, Budapest, Hungary.

KUNOVA, V.
Department of Pediatrics, Charles University, Prague, Czechoslovakia.

LÁSZLÓ, F.A.
Biological Research Center of the Hungarian Academy of Sciences, Szeged, Hungary.

LEISZTNER, L.
Institute of Forensic Science, Budapest, Hungary.

MAGYAR, K.
Department of Pharmacodynamics, Semmelweis University of Medicine, Budapest, Hungary.

MALINOWSKA, I.
Institute of Chemistry, Maria Curie-Sklodowska University, Lublin, Poland.

MATKOVICS, B.
Biological Isotope Laboratory, József A. University, Szeged, Hungary.

MATUS, Z.
Institute of Chemistry, University Medical School, Pécs, Hungary.

MOLNÁR, I.
Human Serobacteriological Works, Gödöllő, Hungary.

MRSKOS, A.
Pediatric Research Institute, Brno, Czechoslovakia.

NAGY, J.
Department of Pharmacology, Semmelweis University of Medicine, Budapest, Hungary.

OHMACHT, R.
Institute of Chemistry, University Medical Scool, Pécs, Hungary.

POSPISIL, R.
Pediatric Research Institute, Brno, Czechoslovakia.

PUNGOR, E.
Department of General and Analytical Chemistry, Technical University of Budapest, Hungary.

PODHRADSKA, O.
Pediatric Research Institute, Brno, Czechoslovakia.

ROZYLO, J. K.
Institute of Chemistry, Maria Curie-Skodlowska University, Lublin, Poland.

SHALABY, A.
Institute of Pharmaceutical Chemistry, Semmelweis University of Medicine, Budapest, Hungary.

SIROKMÁN, F.
Biological Research Center of the Hungarian Academy of Sciences, Szeged, Hungary.

STOURACOVA, O.
Pediatric Research Institute, Brno, Czechoslovakia.

SZABÓ, L.
Biological Isotope Laboratory, József A. University, Szeged, Hungary.

SZÁSZ, Gy.
Institute of Pharmaceutical Chemistry, Semmelweis University of Medicine, Budapest, Hungary.

SZÖKŐ, É.
Department of Pharmacodynamics, Semmelweis University of Medicine, Budapest, Hungary.

TAKÁCS, T.
University of Medicine, Szeged, Hungary.

TÓTH, G.
Department of Medical Chemistry, University of Medicine, Szeged, Hungary.

TÓTH, T.
Department of Applied Chemistry, Eötvös L. University, Budapest, Hungary.

V. TÓTH, M.
Department of Medical Chemistry, University of Medicine, Szeged, Hungary.

TRNKA, V.
Department of Pediatrics, Charles University, Prague, Czechoslovakia.

VAJDA, J.
Institute of Forensic Science, Budapest, Hungary.

VARGA, J.M.
Department of Dermatology, Yale University, Medical Scool, New Haven, Connecticut, USA.

VERESS, G.
Department of General and Analytical Chemistry, Technical University of Budapest, Budapest, Hungary.

VILETOVA, H.
Department of Pediatrics, Charles University, Prague, Czechoslovakia.

INDEX